ORDER OF
BUILDINGS AND CITIES

YAN GU

ORDER OF BUILDINGS AND CITIES

A PARADIGM OF OPEN SYSTEMS EVOLUTION FOR SUSTAINABLE DESIGN

PETER LANG
Bern · Berlin · Bruxelles · Frankfurt am Main · New York · Oxford · Wien

Bibliographic information published by die Deutsche Nationalbibliothek
Die Deutsche Nationalbibliothek lists this publication in the Deutsche Nationalbibliografie; detailed bibliographic data is available on the Internet at ‹http://dnb.d-nb.de›.

British Library Cataloguing-in-Publication Data: A catalogue record for this book is available from The British Library, Great Britain

Library of Congress Cataloging-in-Publication Data

Gu, Yan,
Order of building and cities : a paradigm of open systems evolution for sustainable design / Yan Gu. -- 1st ed.
p. cm.
Includes bibliographical references and index.
ISBN 978-3-03-430621-8
1. City planning--Environmental aspects. 2. Housing--Environmental aspects. 3. Climatic changes. 4. Industrialization. 5. Sustainable urban development. I. Title. II. Title: Paradigm of open systems evolution for sustainable design.
HT166.G8 2011
711'.5--dc23
2011026114

Cover illustration: The author's design project – City like a dream, 2004, Bartlett School of Architecture, University College London, UK.
Cover design: Didier Studer, Peter Lang AG

ISBN 978-3-0343-0621-8

Hochfeldstrasse 32, CH-3012 Bern, Switzerland
info@peterlang.com, www.peterlang.com, www.peterlang.net

Printed in Switzerland

Preface

This book is dedicated to those chasing beauty and poise between sciences and arts, an inquiry from the sociological and ideological perspectives and explored with scientific foundations of complex system science and information science, to generate something novel. It is a manifesto of my native culture. It refers to both ecological philosophy extended in sustainable design and complexity science applied in research methodology, a revival of classical wisdom in the era of post-modernism, post-industrialism, post-capitalism and globalization.

From the philosophical viewpoint of classical Platonism, concerning space, time and man, the sustainability of the planet is hypothesized as a development at a decent rhythm, in a humanly acceptable and environmentally friendly pattern, in order to reach the complexity, richness and diversity and sustainability of human inhabitation for the long term.

The grand mixture of disorder and order in the model of open systems evolution or dissipative structure suggests: we need to be wild to explore the world without edges or boundaries, as the nature of mankind's life to fulfil the sense of being alive; from this perspective, life means chaos, while this chaos brings up order, i.e. the exuberance of life in liberty.

Yan GU
May 2011

Abstract

The assumption of an environmental crisis, e.g. global warming and climate change, has put into question the viability of the modern development pattern, especially in industrialised societies since the Industrial Revolution. This pattern is charactered by excessive exploitation of energy and resources without concern for negative impact upon the natural ecosystem. One of primary challenges to sustainable development is identified as the dilemma between long term economic development and environmental damage, composed of temporal, social, economic, environmental, cultural and technological dimensions.

Influenced by this industrial pattern, modern design of buildings and cities contributes to environmental degradation. To explore an alternative paradigm for sustainable design, a model of open systems evolution is investigated based on the scientific foundations of the Second Law of Thermodynamics and complex systems science. As a manifesto of the Post-Modernism world-view, this new paradigm states that the creativity of the universe appears as the emergence of order or organised complexity via the mechanisms of open systems evolution. An open system spontaneously adapts to its host environment via gradients and uses inputs of energy and resources for its evolution towards order, which is supported by an internal structure through sclf-organisation, minimising negative impact and optimising compatibility with the host environment as measured by entropy.

The paradigm of open systems evolution implies a design framework for sustainable co-existence of man and nature, which directly impacts institutional education and design practice for the built environment. Technically, it suggests an intelligent model of sustainable design, in which, order is interpreted as the sustainable symbiosis of nature, buildings and cities. Theoretically, this proposition of sustainable design is a contextual design strategy for ecological sustainability of buildings and cities, adapting to the natural ecosystem and ensuring positive environmental impact.

Acknowledgements

Gratitude to all those people, for inspiring and stimulating me through the harshness and the toughness of developing the novel ideas, for encouraging and cheering me up with the delight and the excitement of eventual achievements:

Rob Evans, John Frazer, William Moran, Iven Mareels, Greg-Missingham, Justyna Karakiewicz, Bharat Dave, Scott Drake, Robert Crawford, Dominique Hes, Paolo Tombesi, Janis Birkeland, Robin Drogemuller, Martin Betts, Kathi Holt-Damant, Kerrie Bianchi, Steven Moore, Zhu Jian-Fei, Han Sun-Sheng, Eckhart Hertzsch, Angela Hass, Bernard Brown, Ilan Vize, Maggie McCormick, Julie Rudner, Inger Mewburn, Crystal Legacy, Catherine Tate, Chiu Chen-Yu, Wang Hui-Chuan, Jin Xiao-Hua, Ou-Yang Zhi-Xi, Li Feng, Xu Jia, Zhang Lu, Wu Ming, Zhang Yan-Jin, She Ling-Ye, Chen Yu-Shen, Beatriz Maturana, Wiryono Raharjo, John Ting, Thieu Minh Tuan, Viet Trinh Hong, Jane Trewin, Holly Davis, Lorenne Wilks, Sara Pheasant, Christopher Stewardso, Rachel Webster, Stuart Wyithe, Peter Taylor, Mandy Li, Assaf Gurt, Wang Wei, Peter Farrell, Thomas Hanselmann, Ba-Ngu Vo, Andrew Hung, Lars Yencken, Leila Koosh, Natasha Baxter, Tracy Painter, Angie Miles, Howard Gu, Lewis Mayo, Yeung Bick-Har, Bohdan Durnota, Matthew Berrymane, Anne-Marie Grisogono, Enrica Colabella, Giuseppe Marinelli De Marco, Mauro-Palatucci, Gerhard Schmitt, Matthias Kohler, Fabio Gramazio, Colin-Fourier, Quinn Stacpoole, Jorge Raedó, Caroline Schopfer, Lee Jia-Kang, Lv Nan, Yuan Bei, Dong Jing, Tang Yi-Ping, Yao Jie-Wei, Zhu Ming, and my mum and dad in China.

This work is partial of PhD research funded by Faculty of Built Environment and Engineering at Queensland University of Technology, Victoria Research Laboratory, National Information and Communication Technology Australia at University of Melbourne, and Complex Open Systems Research Network in Australian Research Council, Australia.

Table of Contents

Part II The Framework of Open Systems Evolution

Part III The Application of Open Systems Evolution to Sustainability

List of Figures

The author gratefully acknowledges the permission granted to reproduce the copyright material in this book: every effort has been made to trace copyright holders and to obtain their permission for the use of copyright material. The publishers apologize for any errors or omissions and would be grateful if notified of any corrections that should be incorporated in future reprints of this book.

Introduction

Questions

The global debate on sustainable development over the last three decades has put into question the viability of the modern development pattern prevalent primarily in Western societies since the Industrial Revolution. In this pattern, societal progress has focused on economic prosperity by excessive exploring energy and natural resources while overlooking impacts on the natural environment induced by human activities. In this industrial pattern over the last two centuries is the current environmental crisis of consisting of global warming and climate change. This crisis has affected progress in sociological sense, a threat to fundamental social well-being and stability on global and local scales. In brief, the dilemma between the aspiration of long term development and the reality of environmental degradation has become one of the major challenges of sustainable development in the 21st century.

In the building industry, influenced by the industrial pattern, buildings and cities are designed as industrial mechanisms without concern for negative impact upon the natural ecosystem, inevitably contributing to the environmental crisis. In brief, the industrial pattern on the larger scale of social-economic development and on the smaller scale of design in the built environment is unsustainable.

Hypothesis

In a comprehensive sociological picture, the concept of sustainability is defined as a complex system with a full spectrum embracing temporal, social, economic, cultural, environmental, and technological dimensions. To achieve sustainable development, it is essential to develop with an integrated balance of these multiple dimensions. In the built environment, the problem of unsustainable design is represented as inadequate concern for the environmental context in design strategy, for which a more integrated design proposition is expected for a sustainable built environment.

To meet the complexity of sustainable development, there are essentially two options: either to improve the existing industrial pattern, or to replace it with a new pattern, a post-industrial pattern. Both of them would need to extend one-dimension of the industrial pattern to cover the full spectrum of sustainability in multiple dimensions. Focusing on design in the built environment, a hypothesis is no less than an alternative paradigm in architecture and urban design. It would be concerned with the environmental impacts of buildings and cities over their complete life-cycle, effectively reduce the negative environmental impacts of buildings and cities, and even positively restore the ecological resilience of the natural ecosystems.

Methodology

To develop an alternative paradigm for sustainable design in the built environment, a scientific framework for an establishment of sustainable interrelations of buildings and cities with the natural ecosystems is essential. The study of the harmony between man and nature has been a traditional theme of ecological design in architecture and urban design. However, it was explored without an explicit principle that connects nature as an environmental context to buildings and cities.

To bridge this gap, a scientific paradigm of open systems evolution, the Second Law of Thermodynamics to open systems is investigated as an alternative theoretical framework for sustainable design. The Second Law of Thermodynamics traditionally describes the fundamental principles of thermodynamic transfers in a closed system, i.e. transfers of energy, matter and information between the system and the host environment. In this transfer, heat death or entropy irreversibly takes place. However, the thermodynamics of an open system provides a striking interpretation of the generation of entropy in a parallel of the life growth on the earth, which contributes an evolutionary mechanism of open systems, i.e. order emerges out of chaos.

The mechanism of open systems evolution is specified as an open system spontaneously adapting to the host environment due to gradients of temperature, pressure, density, etc, takes use of available resources as inputs from the host environment, such as energy, matter and information, and evolves from a disordered state to an ordered state. At the microscopic level, the open system self-organises a highly ordered internal structure in the direction to an optimal distribution of resources within the open system. The maximum entropy probability principle is one of alternative optimal distributions, i.e. maximizing the distribution of available resources across all agents within the system resulting in an optimal allocation of energy and resource distribution in the open system.

At the macroscopic level, along with this microscopic optimisation is a stable compatibility of thermodynamic interrelation between the open system and its host environment. This is indicated by the minimisation of entropy from an initial far-from-equilibrium state to a final 'non-equilibrium' state, when the open system finishes its evolution and establishes a compatibility with the host environment. This final state is defined as order of open systems evolution.

Objective

According to the paradigm of open systems evolution, a conceptual model of sustainable design in the built environment is constituted in three steps. First, it is essential to establish open thermodynamic relationships of buildings and cities with their natural environment. For example, it can be realised by maximizing available resources, not only the prevalence of unrenewable energy and resources of fossil fuel and electricity, but also the renewable solar, wind, geothermal and etc, in order to produce minimal negative environmental impacts to the natural ecosystem. Second, it is essential to optimise the performance of buildings and cities in the environmental context in the evolutionary mechanism of open systems. This mechanism consists of adaptation, self-organisation and optimisation, for an optimal distribution of energy use and resource consumption in buildings and cities, with minimal negative environmental impacts, e.g. waste and pollution emission. Third, it is necessary to consider the micro-climate demands of the end-users of buildings and cities into the optimisation of environmental performances. The objective of buildings and cities in open systems evolution is the sustainable symbiosis with the natural environment, which is generalised as order of the built environment.

Based on the theoretical paradigm and the conceptual model, an information control system for effectively controlling and adapting environmental performance of buildings and cities is constituted as such: in a virtual information system, buildings and cities as open systems, the natural environment as their host environment, and the end-users of buildings and cities will be parameterized and configured into an integrated scheme for the optimisation of environmental performances, evolving as an intelligent system in the mechanisms of open systems evolution. In summary, sustainable design in the model of open system evolution is a holistic proposition, within which building and cities, man and nature are integrated for an environmental sustainability.

Part I
The Concept of Sustainability

Chapter 1
Un-sustainable development: an ill-posed question or well-defined problem

By critically analysis and identifying the industrial development's foundations of the philosophical anthropocentrism, the motivation in the neo-classical economic regime, consumerism culture; followed by the negative environmental and social impacts; in addition, its limits and constraints such as natural resource depletion and global population increase, the industrial development pattern will be generalised as unsustainable. Particularly, the modernisation of China since the late 20th century will be specified as a contemporary case of the industrialised urbanisation. It will be generalised that development in the industrial pattern is unsustainable at both global and regional scales. Responding to unsustainable development are global campaigns and debates for sustainable development over the past three decades, which has been extensively engaged by social, economic, environmental and scientific groups. It will be summarized that sustainable development is a complex notion, systematically composed by multiple dimensions.

1.1 The industrial development pattern for the modernisation

1.1.1 Industrial development in Anglo-Saxon societies

Urban development in the industrial pattern was originally devised in Britain since the Industrial Revolution from the 18th century to the 19th century, opening the horizon of the modern civilisation. It lately transferred to North America in the grand exploration and colonization in the 19th century for the industrial mass-production and the capitalism in the 20th century, bringing up national economic prosperity, mostly in Anglo-Saxon societies, with the features of deregulation, privatization, labor flexibility and a higher acceptance of risk (Oppenheimer, 2006).

Initially, as promoted by Adam Smith, the father of modern neo-economic model, the industrial development is motivated to improve human capability to effectively exploiting natural resources as one capital, taking use of human resources as another, to realize economic prosperity in a more efficient and radical manner than in agricultural society (McDonough and Braungart, 2002; Steele, 2005). In addition, scientific progresses have been achieved in reductionism, since the use of tools was essentially required and encouraged; technology and technical movements have been dramatically accelerated than ever before (Dobson, 1990; Stasinopoulos, Smith, et al, 2008). All these material progresses subsequently realized the evolution of human civilizations and liberations in intellectual, ideological, cultural, sociological and technological dimensions (Hagan, 2001; McDonough and Braungart, 2002).

In brief, the Industrial Revolution is a radical achievement of mankind transferring from agricultural society to industrial one, one of the most advanced progresses in human history. In consequence, the industrialisation realised the modern urbanisation indexed by modernisation, industrialisation and mobility in Europe and North America earlier than in other continents (Gauzin-Müller and Favet, 2002; Mcdonough and Braungart, 2002; Global Statistics, Research Group on the Global Future, 2005; Salat, 2006).

1.1.2 Anthropocentrism philosophy in the industrial societies

In Western civilisation, the debates on mankind's dialogue with nature can be traced back to the time of Plato. In the ancient world, nature was regarded with awe and fear. Natural disasters, earthquakes, hurricanes, volcanic eruptions could be explained away as expressions of displeasure by the gods, while people were helpless in the face of its power. On the other hand, the ancient Greeks saw the world through a filter of mechanistic analysis and revered geometry as an indication of Man's superiority over nature, which prepared the emergence of anthropocentrism in the industrial world.

After the major scientific discoveries of the 16th century to the 18th century, Western societies, with the industrialised instruments and greed-driven economies, progressed in a belief to conquer nature through science for the benefit of humans (Wines, 2000). Liberated by the scientific progress is an obsession of anthropocentrism dominating in these societies. It is believed man is at the centre of the universe and nature exists for the convenience of humans; therefore, man is in his own omnipotence and ability to draw unrestrainedly on the earth's resources (Gauzin-Müller and Favet, 2002). Legacies to this anthropocentric conceit are an aggressive growth of industrial "conquer-the-earth" mentality and the absolute sovereignty of human over nature. This radical philosophical change replaced the previous earth-centered philosophy, and alienated human from nature in an omission of nature's existence along with human progress in the industrial pattern.

In summary, Western societies since the canonization of anthropocentrism have neither the environmental appreciation, nor the ritualized observances; while nature seems to function as an infinite resource provider for their excessive exploration of energy and natural resources. Furthermore, the dominance of anthropocentrism in Western societies led to a reliance on techno-centrism technocracy, profit-motivated capitalism and consumer-based culture. Nowadays, this anthropocentrism is thwarted by nature's revenge in the form of global warming, climate change, the spread of disease, and other natural catastrophe.

1.1.3 Neo-classical economic model in the industrial societies

It is argued, the industrial development pattern has inherited the neo-classical model of economy derived from Adam Smith in the 18th century (Moore, 2006). In this model, nature was viewed as little other than the source of free raw materials and a place to dump wastes for industrial products, while industrial products usually do not contribute positively to nature in another way, as shown in Fig 1–1. Neither the health of natural systems, nor an awareness of their delicacy, complexity, and interconnectedness, has been part of the industrial design agenda (Birkeland, 2002; McDonough and Braungart, 2002). Humans, by contrast, extract and burn fossil fuels such as coal and petrochemicals that have been deposited deep below the Earth's surface, supplementing them with energy produced through waste-incineration processes and nuclear reactors that create additional problems. They do this with little or no attention to harnessing and maximizing local natural energy flows.

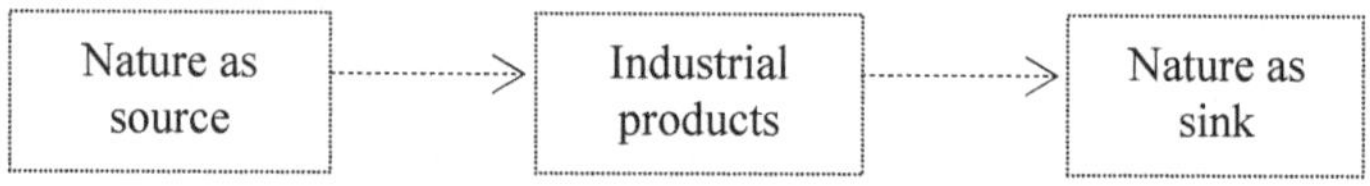

Figure 1–1 The metabolism of industrial production in a linear neo-classical model

However, the industrial pattern in neo-classical economy model is challenged by contemporary economists for the concern with ecological consequences. For example, Daly (1999) argues, that the human economy exists within a larger natural economy; homo economy should be responsible, not just for the initial cost of production, but the cost of products' operation and decommissioning, the impacts upon all those other sites from where materials and energy were mined or will be buried, as shown in Fig. 1–2. To fail to account for the ecological costs of "source", e.g. the natural resources, or the costs of disposing of "waste" in natural "sinks", is only to delude oneself that such costs are external to the economic system. This idea is extended by Hawken, Lovins, et al (1999) as the world's economy being within the larger economy of natural resources and ecosystem services that sustain us. In summary, ecological sense should no longer be missing in the economical system.

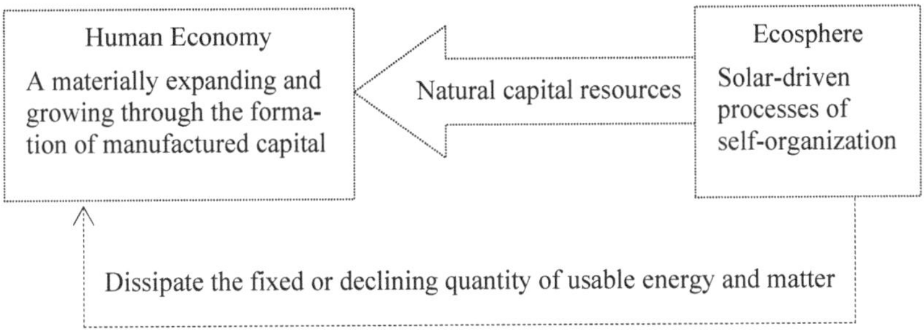

Figure 1–2 Ecological interrelations of homo economy and ecosystem

At the capital expenses of the natural environment, consumerism in the industrialised society has accelerated the activities of economic prosperity and industrial productions. It is predicted (Ray-Jones, 2000), in the next 50 years, global economic activity is expected to increase roughly fivefold, while global manufacturing activity, energy consumption, and the throughput of materials are likely to rise threefold.

1.2 Negative impacts of the industrial development

Since the late 20th century, the industrial development pattern has been delivered to developing countries with the wave of economic globalisation, which seems to bring more widely victors of techno-centrism and anthropocentrism globally. However, the viability of the modern industrialisation development pattern is now under questioned from several perspectives.

First, the industrial development pattern, practised by the industrialized countries over the last two centuries is incompatible to other regional contexts. The unique context of Asia, for example, is charactered by the large populations and hyper-density habitat. It is incomparable by neither the compact model of United Kingdoms with limited geographic resources and small population scale, nor the spreading-all-over model of United States of America with abundant geographic resources but relatively small population scale.

Second, the exacerbating environment crisis such as global warming and climate change, caused by the industrialisation, urbanisation and agricultural activities, indicate this development pattern is unviable in the environmental sense at the global scale, threatening the long term sustainable human habitat on the planet. In addition to the environmental crisis and natural resource depletion at global scale, are cross-national tension among industrialised countries and un-industrialised nations, and intellectual and ethic progresses which have been overlooked in the pursuit of economic prosperity. In the vision of global sustainable development with both environmental and sociological concerns, an indispensable task is to constitute an alternative development pattern, to replace the industrial development pattern, which should be environmentally responsible, contextually compatible and universally applicable.

1.2.1 Negative impacts from environmental perspective

By the early 1990s, the citizenry of every nation began to personally experience unusually severe environmental devastation in the form of heat waves, soil pollution, global droughts, massive crop failures, atomic explosion, oil spills, and the increased incidence of disease. More recently, the limited food and water supplies, lead in gasoline, tropical rain forest, nuclear waste pollution, lethal viruses, and etc, some unprecedented catastrophic event is deemed inevitable. It is the industrial countries who bear the most of the responsibility for global warming and climate change; however, the effects are felt acutely globally including the developing countries.

1.2.1.1 Climate change

Officially, in 2007, the Intergovernmental Panel on Climate Change (IPCC) released its fourth report on climate change (IPCC, 2007), which received much more public attention than previous three ones. Based on scientific, technical and socio-economic information, IPCC (2007) recognizes that current global climate change, i.e. damage in the natural and human environment systems, is anthropogenic and human induced. Globally averaged temperature since the mid of 20th century is very likely due to the observed increase in anthropogenic greenhouse gas concentrations, which include carbon dioxide, CFCs and related gases, methane and nitrous oxide (IPCC, 1990).

1.2.1.2 Global warming

Furthermore, IPCC reports, anthropogenic warming since 1970 is 50 percent above pre-industrial levels. The world has warmed by about 0.74°C in the last 100 years, and will warm a further 0.2°C per decade for the next two decades. Temperatures are likely to rise by about 3°C by 2100, with a range of 2 °C to 4.5 °C. In result, as shown in Fig. 1–3 and Fig. 1–4, the earth's surface temperature increase, and the sea-ice extent decrease.

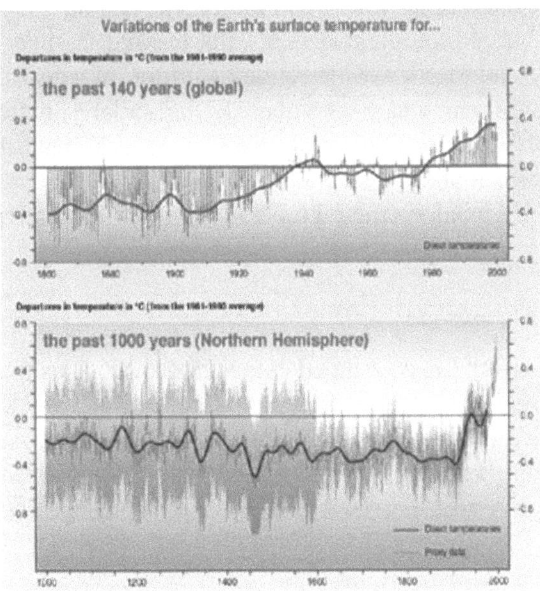

Figure 1–3 Variations in the earth's surface temperature in North hemisphere[1]

1 Diagram source: The IPCC Third Assessment Report: Climate Change 2001, Synthesis Report, Contribution of Work Group I, II, and III to the Third As-

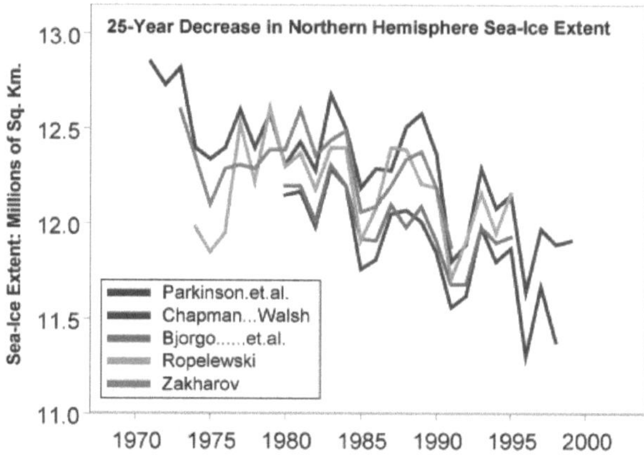

Figure 1–4

This global warming has had a discernible influence on global natural systems (IPCC, 2007), including physical and biological systems, hydrological systems, terrestrial biological systems, marine and freshwater biological systems; in addition, a wide range of future impacts on ecosystems, industry, settlement and society, health. Particularly, there are a range of impacts affecting people's daily lives in the continent of Asia, Australia and New Zealand. It is concluded that the impacts are expected to increase with global average temperature, a risk to human development.

1.2.1.3 Greenhouse gas emission

The reason for global warming is concluded (Mackenzie, 2007) as the stock of greenhouse gases in the atmosphere, including carbon dioxide, CFCs and related gases, methane and nitrous oxide (IPCC, 1990). As shown in Fig. 1–5, the breakdown of world greenhouse-gas emission by energy, agricultural by-products, land use, biomass burning, and

sessment Report of the Intergovernmental Panel on Climate Change. Reproduction of diagram in this book is granted without formal written permission by IPCC, as stated in IPCC Copyright rules.

2 Diagram source: 25 years decrease in Northern Hemisphere Sea-Ice Extent, Fact 5, Poly well: the Five Horseman.

waste, are contributed by human induced energy and resource use in urbanisation, industry and agriculture.

The greenhouse gas emission from the use of coal in the 19th century of Industrial Revolution and the use of fossil fuel in the 20th century (MacKay, 2008) contributed the concentration of carbon dioxide into the atmosphere of biosphere. As shown in Fig. 1–6, it accounts for 55% of the total global warming effect of greenhouse gases. The burning of fossil fuel releases more than 21 billion tons of carbon dioxide into the atmosphere each year (Gauzin-Müler, 2002). The current concentration of carbon dioxide in the atmosphere, 380 parts per million (ppm), having been rising for at least half a century, opposed to the 280 ppm prior to the industrial revolution, is greater than it has been at any time in the last 650,000 years, according to ice cores that contain trapped bubbles of earlier atmosphere.

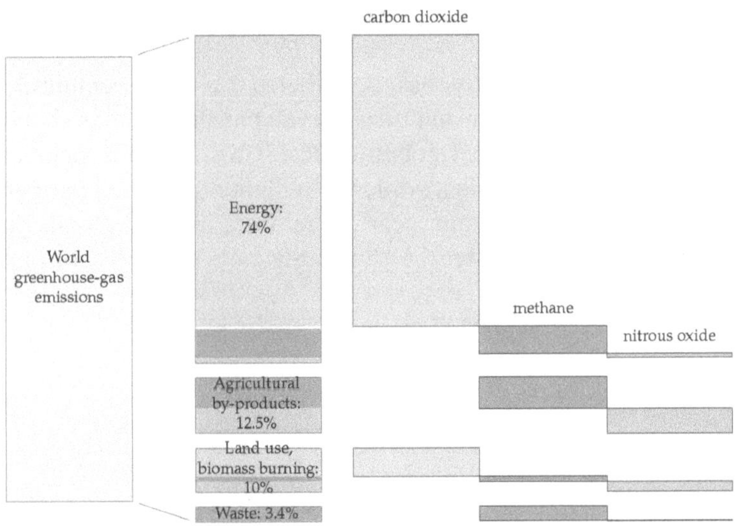

Figure 1–5 Breakdown of world greenhouse-gas emissions by cause and by gas[3]

3 Diagram source: Emission Database for Global Atmospheric Research (MacKay, 2008).

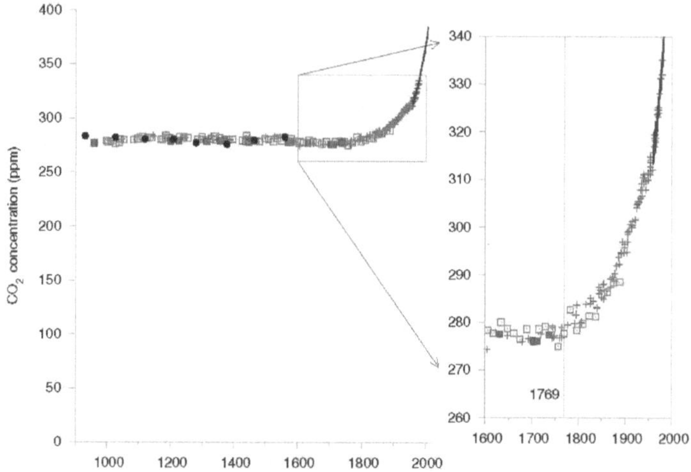

Figure 1–6 Carbon dioxide (CO_2) concentrations for the last 1100 years[4]

Respectively, each country has contributed to the greenhouse gas emission according to their industrial development level. As shown in Fig. 1–7, the industrialised countries like USA has less population than the countries like China, while the pollution produced per person is much higher than the latter, nearly five times larger. In total, China nowadays ranks second after the United State of America contributing to global warming change.

4 Diagram source: Intergovernmental Panel on Climate Change, The IPCC Third Assessment Report: Climate Change 2001, The Scientific Basis, Technical Summary, Figure 8: Records of changes in atmospheric composition (MacKay, 2008)

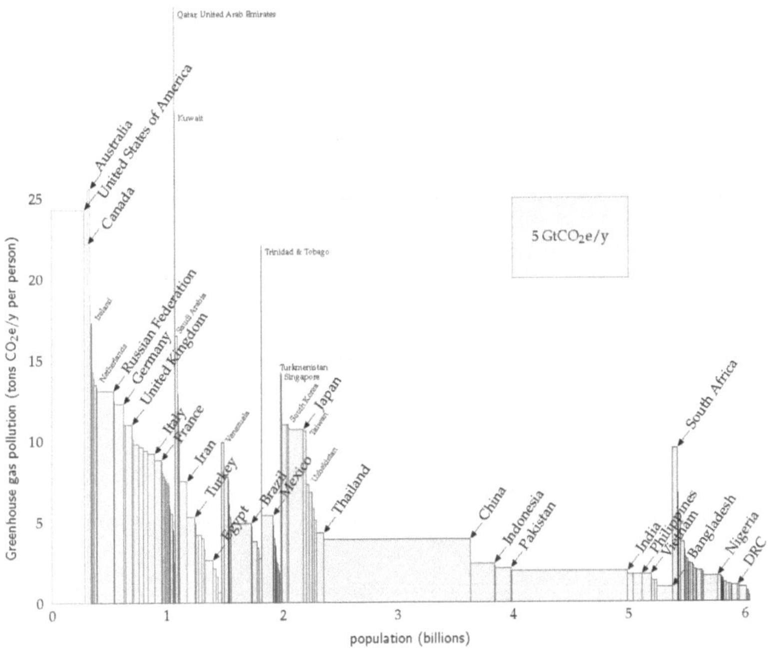

Figure 1–7 Greenhouse gas pollution per person classified by countries[5].

1.2.1.4 Responses to climate change

It is recognized, the dominance and addiction of energy use on fossil fuels has caused the problem of unsustainable environment, as shown in Fig. 1–8. The most efficient strategy to eliminating the environmental problem is to reduce energy demands by population reduction and lifestyle change, and to replace fossil fuels with more efficient and clean renewable energy. The array of potential solutions to climate change which are available to human societies ranges from purely technological, behavioural, managerial and to policy (IPCC, 2007). In conclusion (IPCC, 2007), a systemic and intergraded solution of adaptation taking into account climate, social and economic changes, to

5 (ibid)

reduce vulnerability to future climate change, by enhancing adaptive capacity and increasing resilience, as shown in Fig. 1–8.

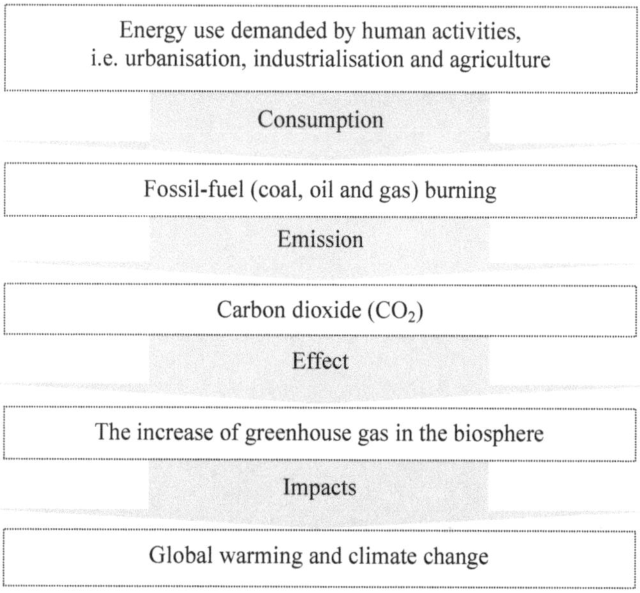

Figure 1–8 The inter-chain from human activities to climate change

1.2.2 Natural resource depletion and scarcity

In addition to climate change, the profligate consumption of energy and natural resources in the industrial development pattern has resulted in the depletion of limited storage of energy and natural resources on the earth. It is argued (Rees and Wackernagel, 1996), industrial development enlarges the ecological footprint through successive demands on natural resources such as water, land and energy; while natural ecosystem has limited productive, assimilative and regenerative capacity for the accelerating human consumption of energy demand and waste. To sustain the population growth and material consumption, the aggregate ecological footprint is already larger than the Earth (Pulselli and Tiezzi, 2009), as shown in Fig 1–9.

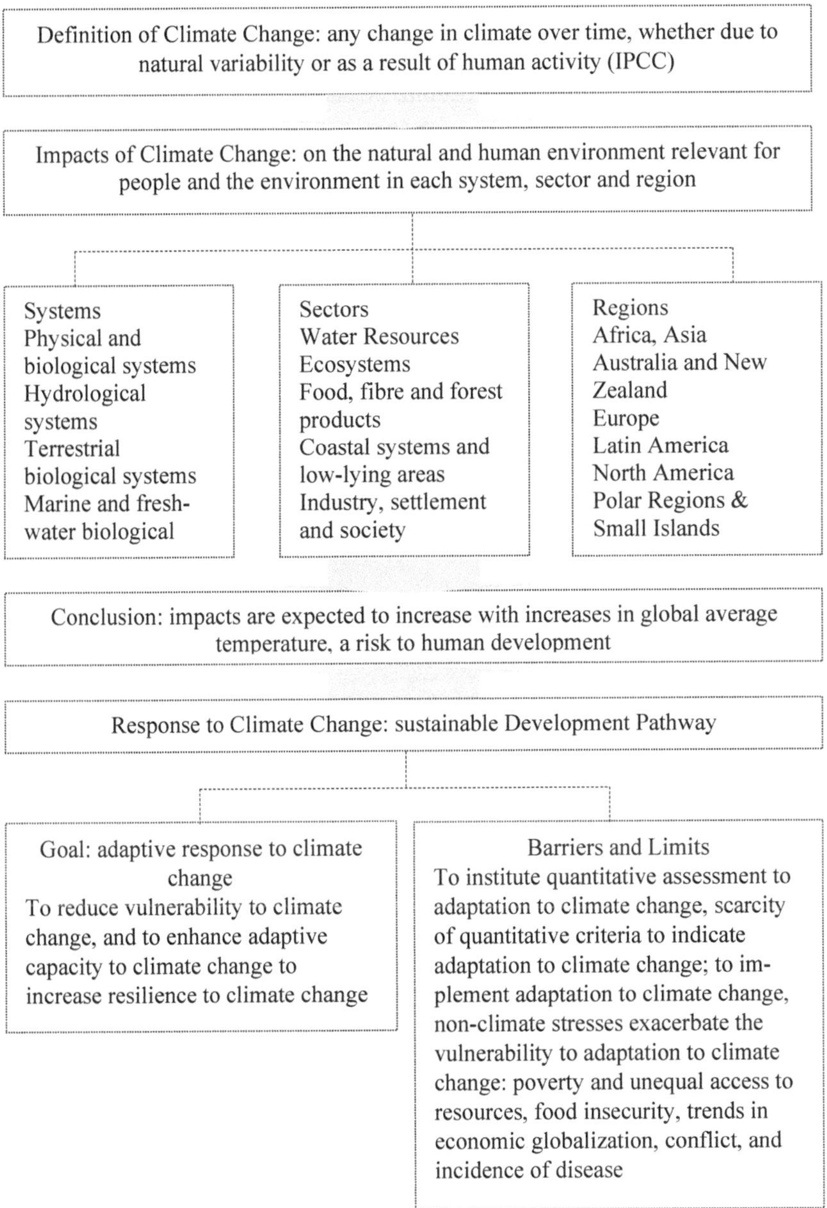

Figure 1–9 Climate change and some responses (IPCC 2007)

Lovelock (2006) argues in earth science, the primary omission of the industrial development is its failure to account for the dependence on the power systems and to the inevitable time when fossil fuel runs out. More complicated relationships of energy consumption with economic gain and urban population growth indicate the environmental price of the industrial development has been comprehensively investigated by Commoner (1972) in environmental science. In brief, such an industrial development cannot continue for much longer without resulting in the collapse of the world's supporting ecosystems (Meadows, Meadows and Randers, 1992; 2005), as shown in Figure 1–10.

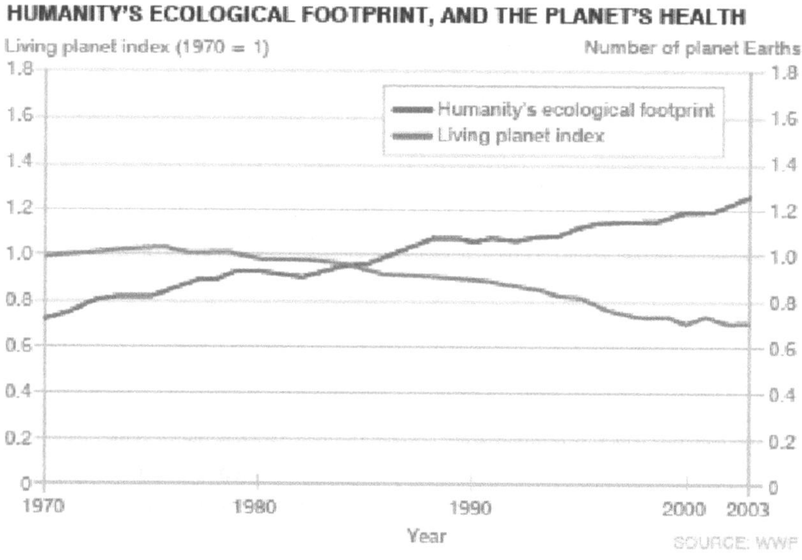

Figure 1–10 Humanity's ecological footprint and the planet's health[6]

6 The global demand for resources exceeds the biological capacity of the earth to renew these resources by some 20% (Date from World Wide Fund for Nature).

1.2.3 Negative impacts from sociological perspectives

In addition to the environmental crisis and natural resource depletion which press stresses on global sustainable development, there are negative sociological and economic impacts, such as the increase of world population and poverty, unequal distribution of wealth, unequal access to available resources of water, food, energy and etc (Behling and Behling, 2000; Hagan, 2001; Birkeland, 2002; Birkeland, 2002; Steele, 2005; Huber and Mills, 2006; Salat, 2006).

1.2.3.1 World population increase

The explosion of global population over the post 100 years will continue increasing in the coming centuries. There are two issues related to the population increase, one is world population increase, another increase of urbanised population. Statistics show that real boom in urbanisation happened at the end of the 1900s (Ehrlich, 1972). In 1900, only around 14% of the world's population, about 200 million people lived in cities. At the beginning of the 21st century, 50% of people world-wide are living in cities; the World Bank predicted that by 2025, 80% of the population will be urbanised, with the total increase up to 5.1 billion by 2050, the bulk of which increases will be in developing countries (Gauzin-Müler, 2002; Salat, 2006). It has been estimated that over the next 40 years, the future urban development will be needed equivalent to a thousand cities, each with 3 million inhabitants, which brings an urgent reality to the need for sustainable principles. In total, the population increase brings up poverty and refugees, unpredicted in human history (Malthus, 1798).

1.2.3.2 Unequal access to global natural resources

Regarding to the drastic tension between developed industrialised countries and developing or even undeveloped countries, people in Europe, for example, nowadays attempt to take full advantages of renewable energy and resources, such as off-shore wind, geothermal and solar energy as "a clean development strategy" to replace the "dirty model practised in industrial nations" (Katz, 2009), while countries in

Asia, Africa and South America are still on the way how to clear up the poverty for people's basic daily survival yet.

1.2.3.3 Unbalanced development on human intellectual growth

Furthermore, as argued by Hungarian architect and professor Friedman (1975) in a discussion of social utopia, the unquantifiable loss of human potential far outweighed measureable gains in productivity, "the social and intellectual growth of the citizen is much more important than the production of goods every ten years." In brief, social justice and non-violence, biological and cultural diversity, democracy and participatory decision making, non-competitive and non-hierarchical forms of social organization, become broadly accepted by the green movement as essential preconditions of sustainable societies (Birkeland, 2002).

1.3 A contemporary case of China's industrial modernisation

Sociologist Castells (1989) articulates a global transfer of industrial production from industrialized countries to developing countries since the early of 1990s. It works as a driving engine for the increasing urbanisation accompanied by intensive industrialization in developing countries across the national boundaries (Steele, 2005). In a contrast to the relative stable and equilibrium urbanisation in developed countries, this transfer imposes drastic social, cultural and economic changes in developing countries, e.g. increase in urban population, building boom, economic flourish, natural resources exploitation and environmental degradation. A conceptual map of global urbanisation spreading from developed industrial countries to developing countries is generalised in Fig. 1–11.

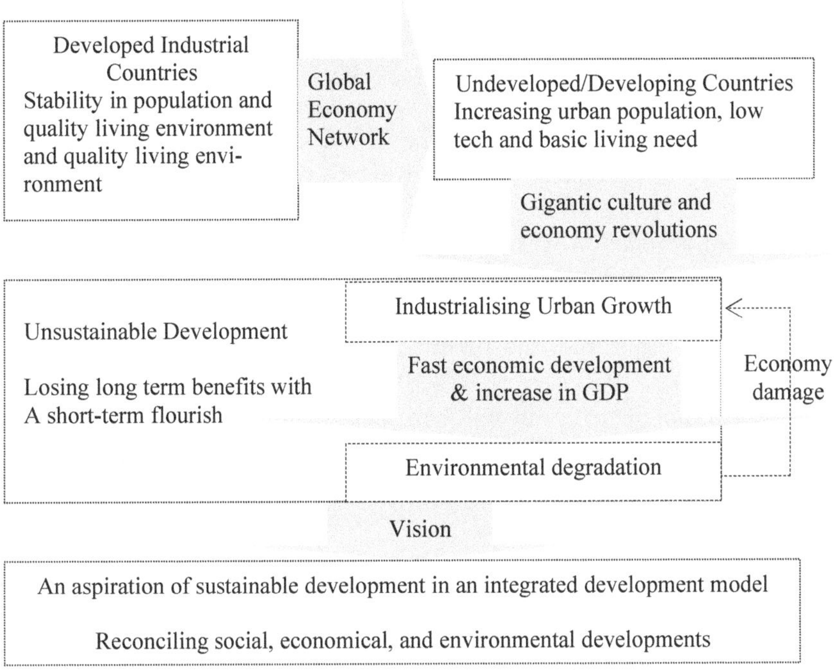

Figure 1–11 A conceptual map of globalisation, urbanisation and industrialisation

In such a globalization backdrop, China's contemporary urbanization in the industrial development pattern has started since 1980's "Open and Reform Policy", spreading from the coastal region, the east of China, to the central and west regions. The national economic boom is witnessed world-wide; China will remain one of main contributor in global economic growth[7]. However, the outlook of urbanization indicated by the increase of urban population and density is complicated, especially by the prices of environmental degradation and natural resources degeneration.

7 Dyer G., China's data put it ahead of Germany, URL: www.FTChinese.com; http://www.ftchinese.com/story.php?lang=en&storyid=001024228

1.3.1 Urban population growth

In the late 19th century, despite the existence of capital cities with populations of one million or more, China's level of urbanization was extremely low, around 6%. The level of urbanization reached 11.2% with the founding of the People's Republic of China During 1950s, the urban population was estimated at 13.6% of the national population, which increased to reach 18% in 1979. The "Reform and Open-door Policies" initiated in 1980s accelerated the process of urbanization. In 25 years, the rate of urbanization doubled, increasing from around 20% in 1980 to over 40% 2000. According to official Chinese estimates, the 50% barrier will be broken through by 2010 (Lü, Rowe and Zhang, 2001). In the globe, there are now 52 countries with an agglomeration of over 2 million inhabitants, and China has 20 such cities.

1.3.2 Urban morphology transformation

Urban growth is justified not only by its population growth, but mainly by the movement of persons and activities towards the outskirts of these large metropolises, which generates the spreading of road network and increased car use (Steele, 2005). To meet such a rapid population growth, urban morphology in China has been changed dramatically, represented by the overspreading of existing cities, the emerging of new cities, and the arising of substantial building programs. The positive implications of this modern urban movement campaign in China lie in two aspects. First, it incredibly changes the country from the state of rural society to urbanised, which arguably improves the citizens' lives in a revolutionary manner. Second, the rapid development of building and construction industry in China has contributed significantly to the national economy (Salat, 2006).

1.3.3 Environmental degradation

However, the achievements of urbanisation in China have been attained at the expense of environmental degradation in the industrial development pattern. By this rapidity of urban growth, ecological footprints of Chinese citizens will move from an average of less than 1 ha per capita to an average as high as 5 and 6 ha per capita, and up to 10 ha capita in cities like Beijing or Shanghai (Rees and Wackernagel, 1996). The environmental vulnerability observed in urban environment of large cities in China, ranges from the affected supply of food, water (Salat, 2006), energy and material resources, the degradation of urban micro-climate, acid rain, the deteriorating quality of indoor and outdoor air, and to obvious health problems. Outside of the cities, the dispersed industrial development patterns produce more traffic and pollution, destroying prime farmland, more erosion and water pollution.

Specifically, two places in China, Linfen in Shanxi province and Tianying in Anhui province have ranked the top ten polluted places in the world[8]. Two of the largest cities in China, Beijing and Shanghai, are recognised as urban air-pollution problems the most severe (Hussey, 1994). Four of China's seven largest river basins are contaminated, and acid rain affects nearly a third of the country (Salat, 2006). All these environmental deteriorations have set China as a big contributor to global climate change. In addition, the actual environmental cost in China has been accounted as a striking proportion of economic production (Johnson, Liu, et al.1997). In brief, in terms of the environmental evaluation of ecological footprint, energy consumption and greenhouse gas emission, China's modern urbanization in the industrial development pattern is recognised unsustainable, a dilemma hindering China's long term's economic prosperity and social stability.

8 The world's most polluted places include Shanxi Linfen and Tianying in China, Sukinda in India, Vapi in India, La Oroya in Peru, Dzerzhinsk in Russia, Norilsk in Russia, Chernobyl in Ukraine, Sumgayit in Azerbaijan, Kabwe in Zambia (Time: www.time.com, 2009).

1.4 A well-defined problem: the industrial development pattern is unsustainable

Since the industrial revolution two centuries ago, human activities of scientific, industrial and commercial developments have progressed drastically, in particular, the 20th century is recognized as an "Age of Progress" (Bisch, 2001), as they reflect on all the opportunities that have come from the explosive growth of industrial production, urban development and economic prosperity. However, as the consequence of the industrial development, the current unsustainable issues raised on the earth ranges from environmental degradation, such as pollution of air and water, the disastrous potential climatic effect of the concentration of greenhouse and ozone-depleting gases in the atmosphere, to energy and natural resource depletion, and population growth and agricultural limits leading to global famine. Thus, the industrial progresses are viewed as the hereditary enemy of true progress by the majority of environmentally sensitized people. Furthermore, anthropocentrism, i.e. humanity holding a privilege of Man's sovereignty in the universe, is criticised having violated its basic ethical responsibility of "each generation's obligation to meet its needs without jeopardizing the prospects for future generations to meet their own needs" (Durning, 1992).

In summary, the modern civilization, having prioritized in the industrial societies over the past two centuries and still colonizes the world cultures, is unsustainable; the so-called civilization progress in the industrial pattern has forged reversely and therefore alienated itself from the integrated prospect of human survival on the earth in a long term; this industrial pattern could no further be implemented to continue the sustainable progress of humanity.

1.5 Global campaign for sustainable development

In response, unequivocally confirmed as quintessentially international problem induced by human activities in the industrial pattern, sustainable movements globally campaign for new policies, strategies, technologies and techniques to restore ecology, foster human well-being and social equity and justice systematically.

1.5.1 Early movements of sustainable development

The energy crisis of the early 1970s, caused when the OPEC countries placed an embargo on oil exports, raised public consciousness of growing scarcity in resources and the forthcoming environmental risk, e.g. greenhouse gas emissions and ozone layer depletion, etc., which led to developments in energy conservation and the consideration of alternative energy sources, such as wide power (Walker, 2006). From then on, there has been a flood of commissions, legislation and international agreements dealing with the issue of sustainability.

Key events include the Brandt Commission, 1977–1983: the Brandt Commission investigated the economic divide between developed and developing countries; the Brundtland Report, Our Common Future, 1987, a direct link between environmental protection and economic development; Rio Earth Summit, Agenda 21, 1992, outlined a program for change that encouraged efficient use of the earth's resources and traditional forms of building; Beijing Declaration, 1995; Habitat II, Istanbul, 1996; Kyoto Protocol, 1996, to achieve more concrete measure, participating nations pledged to bring average greenhouse gas emissions over the period 2008 – 2012 back to 1990 levels; Copenhagen Climate Summit, 2009, attempt to forge a global framework on reducing greenhouse gas emission.

At the grass-rooted scale, the environmental movement was initiated in the late of 1960s by a generation that rejected the excess of the consumer society and called for an end to unrestricted economic growth (Gauzin-Müler, 2002; Steele, 2005). Especially, in the ideological battle between environmentalism and consumerism, classicism, traditional vernacular, humanism and even mysticism, eco-economy in minimalism and dematerialisation are advocated as solutions to mediate the consumerism culture (Hagan, 2001; Birkeland, 2002).

1.5.2 Identification of the notion of sustainable development

Responding to increased environmental awareness and recognition of the fragility of the planet, social inequities and human rights, the term of sustainable development addresses three interrelated areas simultaneously: environmental stewardship, social equality and justice, and economic issues, as shown in Fig 1–12.

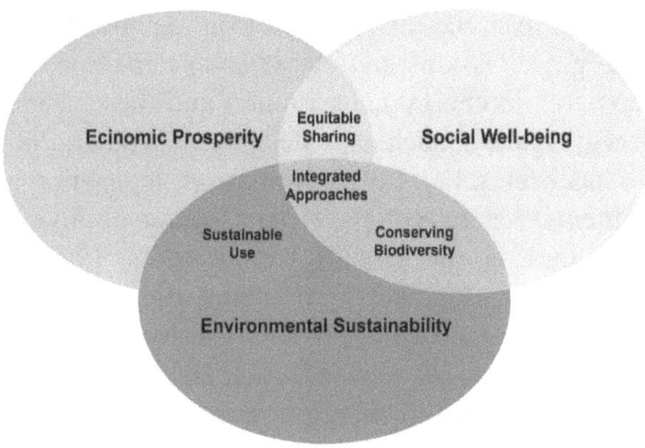

Figure 1–12 An integrated concept of sustainability in economic, social and environmental senses

Formally, the term of sustainability is coined by a series of institutional initiative, primarily guided by the United Nations, characterised as a compromise between the "growth" and "non-growth" factions of the environmental movement in the late 1960s and early 1970s (Steele, 2005). In 1987, an initial official definition of sustainable development was launched in the Brundtland Commission (Brundtland, 1987) which states: "development that meets the needs of the present without compromising the ability of future generations to meet their own needs". At the 1992 Rio Earth Summit, sustainable development is defined as "development which fulfils current needs without compromising the capacity of future generations to fulfil theirs." Both of these definitions emphasise sustainable development as a long term development for the future generation, rather than a short gain for the present generation only. This principle of suitability was enshrined and formulated in the Agenda 21 (Docalavich, 2007), as a common framework of action for all countries to achieved sustainable development in the 21st century.

1.5.3 Environmental reform

The first use of the word "sustainability" in the environment was in a publication in 1980 by the International Union for the Conservation of Nature (UCN) entitled "World Conservation Strategy". Nowadays, many publications on environmental reform in relation to sustainable development range from resource conservation, land use, waste disposal, and health care. However, the main obstacles to progress for a reversal of the environmental catastrophe usually are conservative political and economic agendas, popular ignorance, and environmental programs biased in favor of economic advantage (Hagan, 2001).

For example, the bottom line of environmental reform is always focused on adjusting to accommodate the objectives of profit margins, stock values, interest rates, employment levels, gross national product, and etc, but seldom on considerations of fundamental change in the entire structure of economic values (Hawken, Lovins, et al., 1999) and the retraining of large segments of the work force in preparation for growth industries like waste management and environmental technology. One of successful examples addressing the relationships among economic enterprise, poverty, energy use and environmental repercussions is Schumacher (1973).

1.5.4 Economy reform in low-carbon-energy technology

A new model of eco-effectiveness economy (McDonough and Braungart, 2000) which should produce profits for companies in the business of making profits, while treating people fairly, well and respecting, even celebrating the natural world, is proposed to replace unsustainable neo-classical model in the linear industrial pattern. International government organizations now favour the establishment of a binding worldwide policy where industrial and national economic success can no longer be measured purely on productivity and monetary growth. Instead, progress must be evaluated in terms of ecological impact. In this way, the costs of environmental restoration resulting from the depletion of resources, quantity of pollution produced, and other forms of damage must appear as deductions from profits on all balance sheets as both an immediate deficit and a long-term liability (Wines, 2000). For example, it is argued that the loss of a country's natural resources must be subtracted from the GNP (McDonough and Braungart, 2002; Steele, 2005), when a business have to pay for the loss of things, they have a powerful incentive not to pollute overconsume.

There is an agreement approaching consensus that a successful program of action on climate change must support two objectives, stabilising atmospheric greenhouse gases (GHGs) and maintaining economic growth (Gauzin-Müler, 2002). Achieving both these aims will require a tenfold increase in "carbon productivity", the amount of GDP produced per unit of carbon equivalents (CO2) emitted. This new economy model is named as low-carbon economy. It is calculated, the economic costs of this "carbon revolution" are likely to be managed at some 0.6–1.4 percent of global GDP by 2030. Reducing emissions and maintaining growth implies low-carbon-energy productivity must increase by ten times, which is proposed in two strategies: one is more efficient method on energy use; another is energy resource shift (Huber, & Mills, 2006).

1.5.5 Environmental economics

Since the late of 20th century, environmental economics is developed, to generate everything from studies on natural capital and the proper calculations of income from depletable natural resources to articulate the connection between population and poverty or loans and ecological problems (Steele, 1997). The speciality of environmental economics is to place quantifiable values on each of the myriad parts of the complex ecological web that make up the environment and to manage natural resources. A basic component of environmental economics (Langston and Ding, 2001) concerns the way in which the economics and the environment are interacting to each other, but not separating, as shown in Fig. 1–13. The interdependence is interpreted as the way humans manage the economy produces impacts on the environment and the resultant environmental quality, in turn, impacts on the future performance of the economy. Thus, consumers will automatically embrace radical resources efficiency once they understand that they can reduce consumption "without diminishing the quantity or quality of services that people want" (Hawken, Lovins and Lovins, 1999). In this method, sustainable development represents essentially a value embodying personal aspirations, ideals and concepts of what comprises a benefit for society.

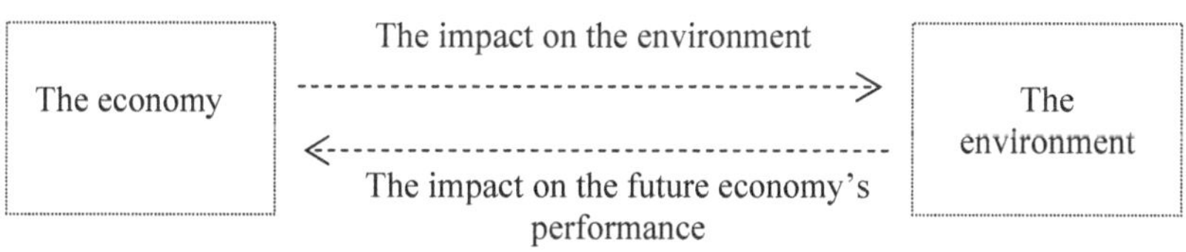

Figure 1–13 Interactions of economy and environment in environmental economics

1.5.6 Scientific propositions on sustainable development

More scientifically and technically, as summoned in the fourth report on climate change by IPCC (2007), sustainable development is to reduce vulnerability to climate change while climate change impedes nations' abilities to achieve sustainable development pathways. The Millennium Development Goals, MDGs 2009 (United Nations, 2009) are one measure of progress towards sustainable development. However, unmitigated climate change would, in the long term, be likely to exceed the capacity of natural, managed and human systems to adapt. Although science to provide policymakers with information about climate change impacts and adaptation potential has improved, it still leaves many important questions unanswered and for further observation and research, for global cooperation on sustainable development summoned, which include (IPCC, 2007):

- Systematic observing and research: investment in research and development in global environmental management, especially focus on efficient use of energy resources and consumption;
- Encouragement of the most efficient possible use of energy resources, and much increased investment in research and development in renewable power and carbon capture and storage, and arrangements for transfer of best technology across the globe;
- This is a huge, long-term and global challenge that involves difficult questions of justice both within and across generations. Humanity's ability to address it is a test of its capacity to manage the consequences of its own actions. So far it has failed. It can afford to do so no longer;
- Discussions between the leading high-income countries and five significant developing countries, i.e. Brazil, China, India, Mexico and South Africa, are essential.

1.5.7 International and national legislations

The international business and government communication are holding a series of conferences to seek global remedies. In 2006, the British government released the Stern Review, a first attempt to quantify the economic costs of climate change. In the same year, the United States Supreme Court ruled that the Environmental Protection Agency has the authority to regulate carbon dioxide and other greenhouse gases. In 2009, three essential elements have been targeted at Copenhagen Climate Summit which aims to constitute a global treaty in legalization (MacKay, 2008), including:

- deep cuts in the CO2 emissions of rich countries alongside curbs on the carbon-intensity of growth in rising economies;
- a financial transfer arrangement to mitigate the cost to poorer nations of a shift to a low-carbon economy; and
- a commitment to embed national targets in a binding international treaty.

However, the target of Copenhagen Climate Summit has arrived in uncertainty, not warranting the concrete investment for reducing the greenhouse gas emission.

1.5.8 Summary

In brief, to maintain and enhance the quality of human life while at the same time conserving natural resources for environmental sustainability will require a fundamental shift in energy strategy, along with strong political will and financial incentives. Sustainable development will depend on the collaboration between community, private institutions and governments, as well as effective utilization and management involved by the public, towards socially and environmental sustainable, fostering the richness of both the cultural and urban fabric (Birkeland, 2002). A conceptual map of the discourse of sustainable development is generalised in Fig. 1–14.

Definition of Sustainable Development
Development that meets the needs of the present without compromising the ability of future generations to meet their own needs (Brundtland)

Impacts of Sustainable Development

Sustainable development can reduce vulnerability to climate change,
Climate change could impede abilities to achieve sustainable development pathways,
Pressures on natural resources and the environment associated with rapid urbanization, industrialization, and economic development,
For developing countries, availability of resources and building adaptive capacity are particularly important.

Problems of Current Sustainable Development
Few plans explicitly included either adapting to climate change impacts or promoting adaptive capacity

Response: urgent actions of global cooperation for sustainable development
Efficient possible use of energy resources
Increased investment in research and development in renewable power and carbon capture and storage
Arrangements for transfer of best technology across the globe

Figure 1–14 Sustainable development in response to climate change (IPCC 2007)

1.6 An integrated proposition of sustainable development

Physicist Jaynes (2003) argues in decision theory, failure to judge one's own loss function correctly is one of the major dangers that human face; and the state of nature tomorrow might be influenced by our decision today. As generalised above, a growing recognition of sustainable development (Singleton, 2004) is that long term economic and social vitality depends upon more efficient, effective use of natural resources, coupled with improved human and environmental health. The challenge is how to sustain and transfer the development from the past, through the present, to the future, with the recognised environmental problems threatening the human survival. In other words, the most complex and problematic issues facing sustainable development is how to construct a human habitat in harmony with the natural environment while meeting the demands of social and economic developments in a long term. This recognition identifies four aspects of sustainable development, as shown in Fig 1–15:

- The scope of sustainable development is a long term course, rather than short one, which implies time as a significant dimension in sustainable development agenda.
- The necessity of reasonable use of natural resources, to meet the demands in economic and social dimensions.
- Both the natural environment and human well-being are affected by the use of natural resources, as essential dimensions of sustainability.
- Nature works as the provider of resources on one hand, and receives the impact of the resources use on the other hand.

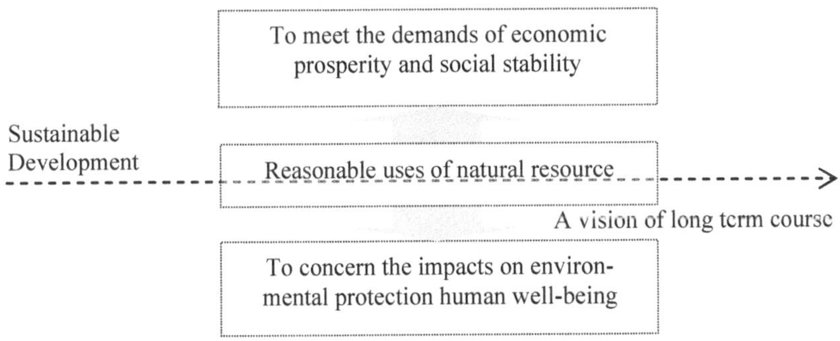

Figure 1–15 Sustainability transfer in a long time course

1.6.1 Sustainable development as a complex system

In brief, sustainable development is an integrated and complex spectrum, which is composed comprehensively by temporary, social, economic, environmental and human dimensions, as formulated in Fig. 1–16. Confronting the current environmental crisis, a deteriorated environment would be unable to support the social and economic demands and to sustain human wellbeing. Thus, the natural environment is recognised as the centre theme of sustainability with the highest priority.

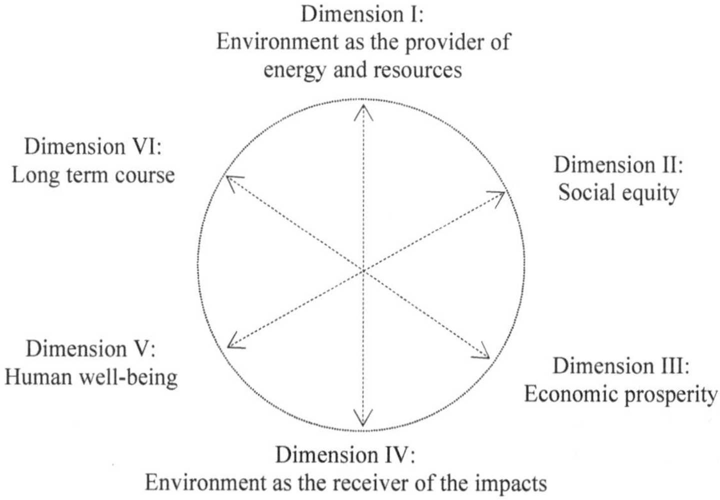

Figure 1–16 Full dimensions of sustainable development as a complex system

1.6.2 Dynamic interactions of multiple-dimension sustainability

Based on the investigation and critical discourse above, it is generalised, sustainability is a dynamic process affecting all aspects of human life, which should resolve the conflicts between the various competing goals, involve the simultaneous pursuit of economic prosperity, environmental quality and social equity as three dimensions or triple bottom lines with the resultant vector being technology (UNESCO, 2001; Voinov and Smith, 2002; Pearce, 2006; Hasna, 2007), as shown in Fig. 1–17. In other words, a comprehensive spectrum of sustainability is composed by temporal, social, economic, environmental, cultural and technological dimensions. Among these complex interactions, the interrelation between economics and ecology of the natural system supported by relative technologies, for example, has been studied in environmental economics and natural resource economics.

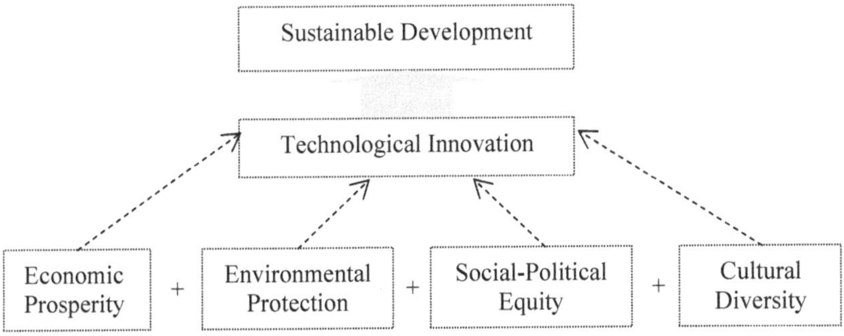

Figure 1–17 Complex interactions in the spectrum of sustainability

1.6.3 Technological innovations to meet the challenges of sustainability

Technological innovation towards sustainability (Allenby and Graedel, 1999) aims at designing, purchasing, or adapting facilities in an environmentally responsible principles, e.g. minimizing consumption, maintaining human satisfaction and minimising negative environmental impacts, which must be evaluated in terms of the materials being constructed, used, refurbished and reused. For example, the Environmentally Responsible Facility (ERF) Matrix is developed to encompass all stages of facility operations and all relevant environmental concerns, corresponding to the life-cycle of a generic facility. Another technological breakthrough towards sustainability is the constitution of Environmental Sustainability Index (ESI), designed by Yale Centre for Environmental Law and Policy and Centre for International Earth Science Information Network, to assess environmental sustainability in countries with integrated criteria of environmental, human and social components[9]. Life cycle assessment is another sus-

9 The ESI tracks relative success for each country in five core components: environmental systems, reducing stresses, reducing human vulnerability, social and

tainable technical tool for environmentally-conscious design, regenerative technology and industrial ecology.

1.6.4 Summary

Since the Industrial Revolution, the ideologies of techno-centrism and anthropocentrism have been fulfilled in a radical manner, to advocate the economic prosperity by excessively exploring energy and natural resources while overlooking the negative impacts upon the natural ecosystem, abandoning other societal imperatives. The complexity of sustainable development requires a systemic approach and trans-disciplinary knowledge to reconcile ecological, economic, social and other implications for a long term achievement instead of a short term gain. Hence, the concept of sustainable development is formulated multiply with temporal, social, economic, environmental, cultural and technological dimensions. An alternative sustainable development pattern, to replace the unsustainable industrial pattern, is proposed to meet the complexity of sustainability, a leap departure from the previous economic dimension only to the multiple dimensions, to mitigate and adapt to the environmental crisis and to enhance the quality of human inhabitation.

---

institutional capacity and global stewardship. In turn, these 5 components form a total of 21 indicators, with each indicator combining two to eight variables, for a total of 76 underlying data sets (International Earth Science Information Network, 2005).

Chapter 2
Sustainable design in the built environment

Argued by scientific historian and philosopher Kuhn (1970), a crisis is always an appropriate prelude to the emergence of new theories. The environmental crisis of the degradation of the natural environment and the deterioration of energy and natural resources in the industrial development is argued as a design crisis, which is induced by the irresponsibility of designers in the built environment concerning energy and natural resource use associated with construction technology (Ryn and Cowan, 1996; Wine, 2000).

Modern cities since the Industrial Revolution in the 19th century consist of human inhabitants, means of industrial production and commercial activities, transport infrastructure of roads and rail, energy infrastructure, communications infrastructure, open space and other public facilities. In modern urban design and planning with features of rationalism, functionalism and mechanicalism, the diverse demands of urban compositional facilities are met by substantial dependences and consumptions of energy and natural resources. In consequence, cities are operated like combustion engines, producing great amounts of pollution emissions and leading to deteriorating liveability on an urban scale. Not only cities developed in this industrial pattern, but also modern architecture in the Industrial Age has been conventionally designed with environmentally de-contextualisation. In brief, buildings and cities are designed without environmental awareness, exploiting abundant energy and material resources from the natural ecosystem, and producing negative environmental impacts upon nature and human well-being. In brief, architecture and urban design in the industrial pattern is environmentally unsustainable.

2.1 Critical discourses on modernism design in the Industrial Age

A series of critical discourses on modernism architecture and urban design in the Industrial Age will be processed in the following sections. Embedded in sociological and scientific backgrounds such as industrial technology and reductionism science, the aesthetic features of modernism design are abstraction, functionalism, mechanism and etc. However, constrained and limited by the backgrounds, modernism design has failed inadequately meet the complexity of other societal imperatives, the most significant of which is the overlooked environmental performance of buildings and cities in the context of natural ecosystem.

2.1.1 Implications of architecture in sociology and culture

Historically, architecture expresses passion, religion and noble thoughts, imposing political, spiritual, ecological and sociological implications; it is one of the most dependable reflections of civilizations' philosophies, the most visible chroniclers of progress, the propaganda vehicles of cultural contexts; its iconography or imagery, in languages varying from industrial, vernacular to classical, is a full integration of social structure, governmental policy, religious rituals, people's relationship to the environment, and philosophic motivations inspired and sustained by the contextual ideology (Jencks, 1997; Wines, 2000; Guy and Moore, 2005). The fundamental architectural concept lies among art, science and technology; it should be considered not only in artistic-aesthetic category, but also in technical-constructive category. For example, the technical-scientific ordering systems that had largely organized the disciplines from which the logic of Modernism had emerged are sociology, the natural science, information and media technologies (Gleiniger, 2008).

2.1.2 *Contexts of modernism design in the Industrial Age*

The historical background of modernism architecture is the Industry Age from the late of 19th century to the 20th century, which is character by anthropocentrism in a mechanistic view of humanity's superiority conquering over nature, neo-classical economy model, industrial technology, and scientific progress in reductionism (Mcdonough and Braungart, 2002). In 1923, the great Modernist pioneer Le Corbusier hailed a "new epoch" and proclaimed: "there exists a new spirit, Industry, overwhelming us like a flood which rolls toward its destined end, has furnished us with new tools adapted to this new epoch, animated by the new spirit" (Erve, 1951). The scientific dominance over the last two centuries was Newtonian and Cartesian science, the features of which consists of reductionism, mechanism and the linear determinism, predictability programme, in progress of simplicity (Prigogine and Stengers, 1984; Jencks, 1997; Stasinopoulos and Smith, et al. 2008; Bellut, 2008; Vrachliotis, 2008). In modern science, the materiality of nature, representing disorder, chaos, unpredictability, was ignored and devalued (Hagan, 2001). In addition, in the model of neo-classical economy, in which nature was viewed as little other than the source of free raw materials and a place to dump wastes, concern for the natural environment was not in priority in the modernism architecture. For example, the tactic of the quality, precision and level of technological advancement are especially explicit in German architecture in the beginning of the 20th century, effectively conveying the message of economic abyss (Steele, 2005).

2.1.3 *Aesthetic features of modernism design*

Within these industrialised social, economic and scientific backgrounds, modernism architecture and urban design are defined as an independent and intervention object in its context, a manifesto of techno-centrism and anthropocentrism, and a celebration of a progress in industry and technology. In the orientation toward a scientific model, the 20th century's architectural Modernism set into motion in a rational program of simplification and objectification (Vrachliotis,

2008). Its aesthetic features include abstraction, functionalism, mechanism, and technically expressive etc.

The logic of Modernism shares a fascination for abstraction and minimalism, in the aesthetic criteria of pure and simple, inspired by both abstract art and reductionism in modern science (Ryn and Cown, 1996; Nicolis and Prigogine, 1989; Wines, 2000b; Hagan, 2001; Gleiniger, 2008; Mainzer, 2008). Le Corbusier is generally associated with the machine aesthetic of purism (Steele, 2005); the legendary slogan "less is more", articulated by Mies van der Rohe in the programmatic ban on ornamentation to the geometric abstraction of primary forms (Jencks, 1997). The language of architecture is dominant by standards, types, and norms (Gleiniger, 2008). Perceivable detail is missing in the modernism design in the Industrial Age, in contrast to Medieval and Renaissance architecture (Wines, 2000b).

In the modernism, buildings and cities were designed as a machine for function and utility in mechanical logic (Portoghesi, 2000; Salat, 2006; Mainzer, 2008); city as a machine for living and working, buildings as machines for living, an equation between a combustion engine and a spiritual vision for shelter, committed to shape-making, space-making in formalist and functionalist invention (Wines, 2000). Le Corbusier is one of the most eloquent celebrants of the rationality of the machine. Particularly since 1970s, the progress of industrial technological advantages was synonymously expressed in high-tech features, i.e. exposed structural systems, vast expanses of plate glass, and cantilevered, tilted, or skewed steel trusses (Wines, 2000).

2.1.4 Inadequacy of modern design

2.1.4.1 Simplicity and monotony

The attempt to impose universal design solutions, especially the technological obsessions in modernism architecture, on an infinite number of local conditions and customs is one manifestation of the modernism principle and its underlying assumption (Ray-Jones, 2000; McDonough and Braungart, 2002). However, it overlooks spatial, natural and cultural diversities in the integrated picture of artistic aesthetics, societal change, human psychology and the growing significance of eco-psychology (Ryn and Cown, 1996; Wines, 2000; Birkeland, 2002). It results in less variety and greater monotony and homogeneity, which are to be read in the metaphors of the collaboration of Mies van der Rohe with the Nazis, the compromise of Le Corbusier with Mussolini and Petain, and of Terragni with the Fascists (Jencks, 1997).

At urban design scale, sociologist Herbert Gans (1962) argued that the wealth and power of any large city was partly dependent on a series of 'urban villages' that were hidden from view: the complexity of different ways of life, ethnic groups, and skilled minorities that were obscured by the dominant culture. Jane Jacobs (1972) attacked the modernist notion that a city could be divided functionally into purified parts, the zoning ideas of Le Corbusier that had helped destroy so many great American cities.

2.1.4.2 De-contextual

As pointed out by architectural theorist Venturi (1950), the Classical concept of architecture does recognise context within composition, i.e. the building as a system of relationships of geometric shapes; it is context gives a building's expression, to which the whole composition of building may possess different degrees of articulation. However, in its concept of universality lack of emphasis on natural and architectural setting, the Classical approach disregards context from an organic standpoint, which results in narcissism in the environmental sense (Wines, 2000).

Nevertheless, Venturi (1950) further argued, one of contemporary problems in architecture dating back to the Beaux Arts Institute of Design is that the building could be designed only for itself, which frequently lacked of indications of buildings' setting, or at best indicated merely the physical dimensions of the site. More drastically, the International Style involving de-contextualised abstraction which is independent of location was the basis of strategies to unify architecture and urban planning in the Modern Age (Gleiniger, 2008). In brief, modernism abstracts is contextual relationships.

2.1.4.3 Environmentally irresponsive

Recently, it is recognised the shortage of environmental awareness in modernism architecture and urban design is the reason for the environmental degeneration in the 20th century. Observed in cities around the world is the absence of environmental commitment in architecture, while associated with industrial imagery and technically driven design style in functionalism and formalist to meet the demand of utilitarian and economic imperatives of modernity (Jencks, 1997; Hagan, 2001). Furthermore, the adoption of wasteful technology by using ecologically offensive materials and the excessive dependence of fossil fuel are ecologically irresponsible, which have caused incalculable environmental damage (Wines, 2000; Gauzin-Müler, 2002). The following sections will specify an environmental discourse of modernism architecture and urban design.

2.1.5 Critical review of modernism design from environmental perspective

2.1.5.1 Urban design in the industrial pattern

In the modern age, one of the most typical urban organisations is the Garden City proposed by Howard in 1898, as shown in Fig 2–1; it was followed by other cases of sprawling Broad-acres City and high-rise based Radiant City (Steele, 2005). Although these urban design programs were originally in a novel approach in a social responsibility to meet the demand for social progress hinged upon material progress. It, in fact, was realised by following the neo-classical economic model, i.e. consuming huge amounts of energy and natural resources from the natural ecosystem as a source, emitting surplus and waste from the consumptions into the natural ecosystem as a sink, rarely acknowledged the environmental consequences of urban development (Ingersoll, 1996; Hagan, 2001; United Nations Human Settlement Programme, 2004). As generalised: "cities have become parasites in the landscape" (Rogers and Gumuchdjian, 1998).

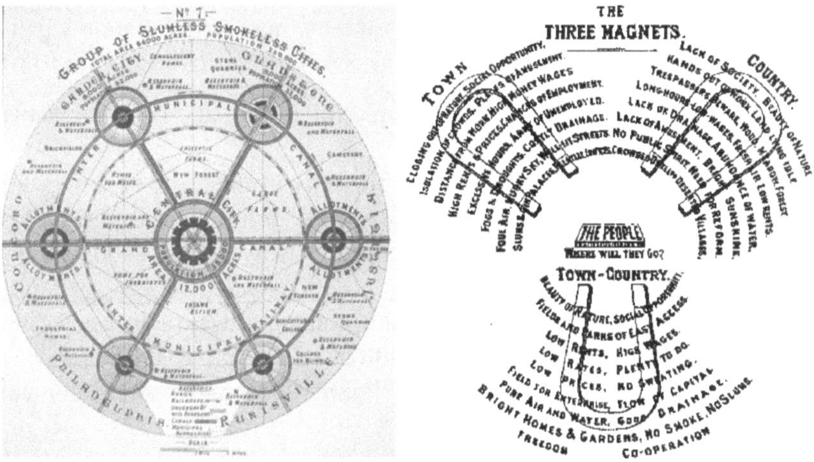

Figure 2–1 Diagrams of Garden Cities proposed by Howard in 1898[10]

In the context of natural ecosystem, the ecological dependences and environmental impacts of cities are termed as urban metabolism (Birkeland, 2002), which has generated environmental impacts on the overall environmental quality of cities and the health of city dwellers. For example, the ecological dependences of cities on the natural ecosystem are classified as follows (Salat, 2006):

– Cities need forest and plants to produce and stabilize earth, to control water reserves by maintaining the catchments' basins through the recuperation of surface water, and to overcome extreme conditions;
– Cities need fertile soil able to eliminate pollutants;
– Cities need a natural absorption cycle to recycle and reduce its waste.

To meet the demands of all the industrial urban facilities, energy and material resources flows are consumed, and pollution emissions are

10 Diagram source: Howard, E., Garden cities of tomorrow (1965), with preface by F.J. Osborn ; with an introductory essay by Lewis Mumford. Cambridge, Mass.: M.I.T. Press. Copyright permission of diagrams is granted for the reproduction in this book by The MIT Press.

thus produced. The environmental problems in urban climate are generalised as follows:

- Over-consumption of energy, depletion of rare and non-renewable materials (Birkeland, 2002);
- Accelerated degradation of urban environment: worsening air quality, heat island, greenhouse effect, acidification and photochemical smog, ozone layer depletion (Ray-Jones, 2000; Birkeland, 2002);
- Noise pollution generated by the giant highways, an escalating amount of solid waste (Salat, 2006);
- The reduction of biodiversity, failure to achieve an ecological balance (Salat, 2006).

In addition to the impact upon the natural ecosystems and environmental degradation of the urban environment by urbanisation, the competitions from agriculture and industry sectors exacerbate the pressure on the demands of the limited energy and natural resources in natural ecosystem, and contributing to climate change and global warming (Ingersoll, 1996; Moore, 2006; Salat, 2006).

2.1.5.2 Architectural design in the industrial pattern

In the Industrial Age, buildings are also conventionally designed in the linear industrial pattern influenced by the neo-classical model (Moore, 2006), profligately consuming energy and resources as the input from the natural eco-system, consequently producing considerable waste to the natural ecosystem. Generally speaking, a building's ecological dependence is composed by the energy and natural resources use from the natural ecosystem, the amount of which depends on buildings' position, shape, structure, materials and energy needs. Two major requirements of energy and natural resources in buildings are generalised as follows:

- The construction of buildings such as the production of building materials by exploiting energy and natural recourses of water, materials, land, wood and fossil fuel (Gauzin-Müler, 2002);

- The operation of mechanistic services for the micro-climatic control within buildings (Stasinopoulos, Smith, et al., 2008), by
- the passive use of energy and resource use, i.e. sunlight, air circulation, landscaping, water and materials; and
- the positive use of electrical energy and other energy for lighting, ventilating, heating and air conditioning, HVAC system (Knowles, 2006).

The construction and operation of buildings represent 50% of natural resources and 40% of energy consumption in Europe (Gauzin-Müller and Favet, 2002), 48% of energy consumption in USA (U.S. Energy Information Administration), and produce 30% of greenhouse-gas emissions. In Australia, the construction of buildings uses roughly 20% of total energy, and operating them uses another 20% (Crawford and Treloar, 2005). Compared to transport infrastructures and other industrial activities in urban systems, the consumption of energy and resources, and waste impacts produced by the building sector are enormous, as shown in Fig. 2–2.

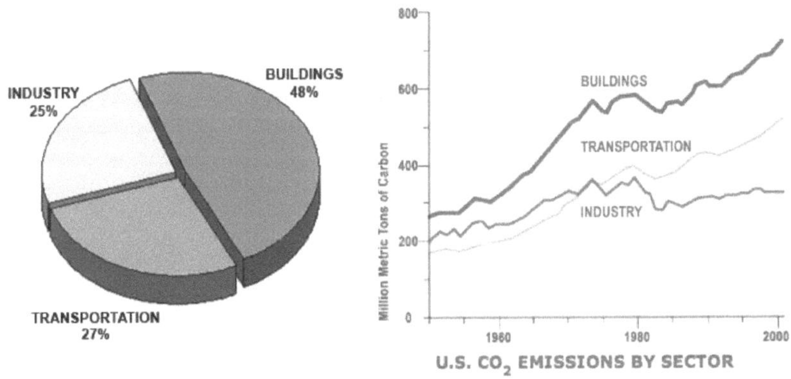

Figure 2–2 US energy and CO_2 emission breakdown by industrial use, transport and building sectors[11]

11 Sources: U.S. Energy Information Administration. Copyright permission of data and diagram is granted for the reproduction in this book by the US Energy Information Administration.

The complex fluxes of energy and natural resources in a building, as shown in Fig. 2–2, start from raw materials, processing into primary energy, converting into secondary forms of energy such as electricity or district heating, continuing in the form of energy used by the end-consumers as supplied energy, then distributed into domestic technical installations as usable energy as shown in Fig. 2–3, and finally a variety of grey energy. In brief, over the life-cycle of buildings, energy and material consumptions start from the fabrication and transportation of materials, the construction on site, the operation of heating, air conditioning, hot water, lighting and power to appliances, the final demolition and removal of rubble, as shown in Fig. 2–4 and Fig. 2–5.

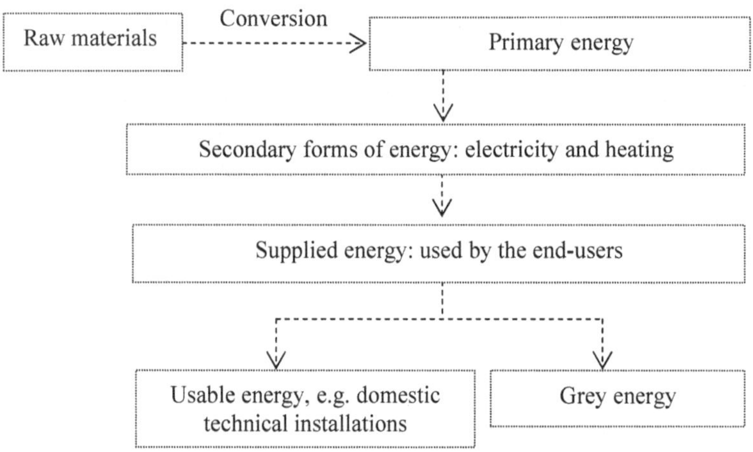

Figure 2–3 Energy and materials flows through buildings

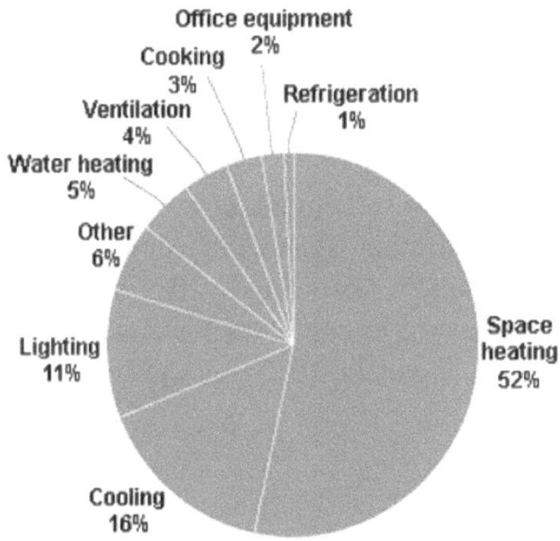

Figure 2–4 Breakdown of end use energy consumption in buildings[12]

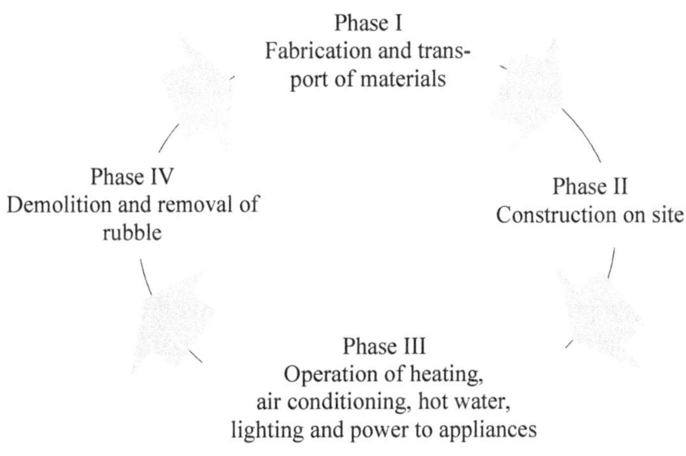

Figure 2–5 Phases of energy consumption in the life-cycle of buildings

12 (ibid)

In buildings, the inefficient electrical and climate conditioning systems squander great amounts of energy; combustion of fossil fuels on-site and at power plants adds greenhouse gas emission, acid rain and other pollutants to the natural environment. In addition, many building materials, furnishings, and finishes give off toxic by-products contributing to indoor air pollution, which results in the effects of causing a large number of diseases, "sick building syndrome" detrimental to occupant's well-being and public health (Birkeland, 2002); and poorly designed lighting and ventilation systems can induce headaches and fatigue (Sustainable architecture white papers, 2000), as shown in Fig. 2–6.

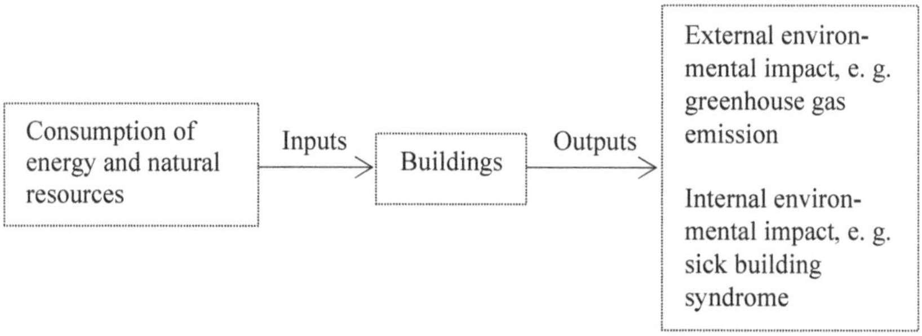

Figure 2–6 The metabolism of buildings in the linear industrial pattern

2.1.5.3 Life-cycles environmental performances of buildings and cities in the industrial pattern

In general, modern buildings and cities, designed in the industrial pattern influenced by the neo-classical model. Over their life-cycles, both resources consumption and waste emission into the natural ecosystems constitute the fundamental interactions or metabolism of buildings and cities with the natural ecosystem. These interrelations are defined as buildings and cities' environmental performances in the context of the natural ecosystem.

Modern buildings and cities, which aggressively attack against nature, taking place all over the civilized world on a daily basis, are now quantified in the method of life-cycle analysis. Argued by Azapagic (2004) in the method of life-cycle assessment, the boundary of life-cycle of a product is not from "cradle to gate", but from "cradle to grave", as shown in Fig. 2–7. Applying this "cradle to grave" analysis to buildings, the life-cycle of buildings starts from planning, design, construction, operation and maintenance, and finishes at demolition. Over the life-cycle of buildings, ecological dependence on the natural ecosystem and waste emission to the natural environment are categorized by Roodman and Lessen (1995), for example, as shown in Tab 2–1.

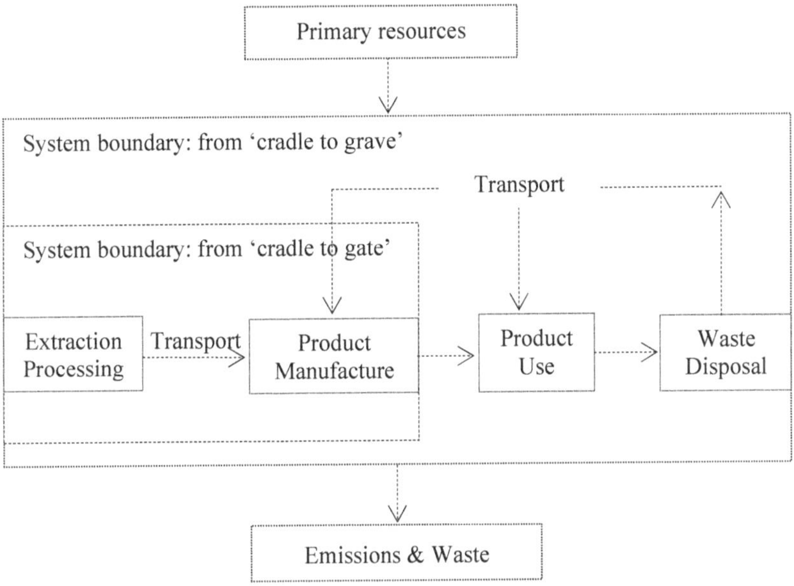

Figure 2–7 Boundaries of life-cycle analysis: 'cradle to gate' and 'cradle to grave'

Table 2–1 The environmental impacts of buildings over the life-cycle

Part I: Consumptions of energy and resources		Effects on environment and human
Use of virgin minerals	40% raw stone, gravel and sand, comparable share of other processed materials such as steel	Landscape destruction, toxic run-off from mines and tailings, deforestation, air and water pollution from processing
Use of virgin wood	Buildings account for one quarter of the world's wood harvest and over 50% of wood is used in the building industry, 25% for construction	Deforestation, flooding, siltation, biological and cultural diversity losses

Use of energy and resources	40% of total energy use, i.e. the total energy consumed in building operation, construction and services in UK is 66% of annual energy consumption, 20% in construction (Vale and Vale 1994)	Local air pollution, acid rain, damming of rivers, nuclear waste, risk of global warming
Use of water	Buildings consume 16% of total water withdrawals, one sixth of fresh water supplies (Brown 1996)	Water pollution; competes with agriculture and ecosystems for water
Landfill	Building waste accounts for 44% of landfill	50% of packaging waste in industrial nations
Part II: Other emission, waste and other negative impacts		
Production of waste	Comparable in industrial countries to municipal solid waste generation, landfill problems, such as leaching of heavy metals and water pollution.	
Carbon dioxide	In the past 100 years, the level of carbon dioxide in the atmosphere has risen 27%; 1/4 of this is attributable to burning fossil fuels to provide energy for existing buildings. Energy use in buildings in the UK is 48% of total CO_2 emissions (Pout 1994).	
Greenhouse gas emission	Buildings account for one third to one half of total greenhouse gases emitted by industrialized countries each year	
Unhealthy indoor	Poor air quality in 30% of new and renovated buildings; higher incidence of sickness, lost productivity in tens of billions annually	

2.1.6 Summary

Despite all the progressive features, such as aesthetics and technologies developed in the manner of reductionism, simplified into functionalism and utility mechanism, which has been institutionalized to be dominant over the last century of the Modern Industrial Age, modernism architecture and urban design have overlooked other imperatives embedded in the complexity of social, economic, cultural, environmental contexts, as shown in a conceptual map Fig. 2–8. It is criticised (Venturi, 1988) as being incapable of adequately conceptualising the diversity, the ruptures and the dynamism as of the urban environment in the Post-Industrial and Post-Modernism Age. In brief, modern architecture and urban design in this linear industrial pattern are environmentally unsustainable.

Therefore, it is necessary to develop the inadequacy of Modernism architecture and urban design by conceptualizing the genuine complexity of human beings and their environment, particularly, the challenges of the global environmental crisis, depleted natural resources and increasing world population. Sustainable architecture and urban design, hence, is embraced with environmental concern with the highest priority over other issues, to alleviate environmental problem and to restore the natural ecosystem.

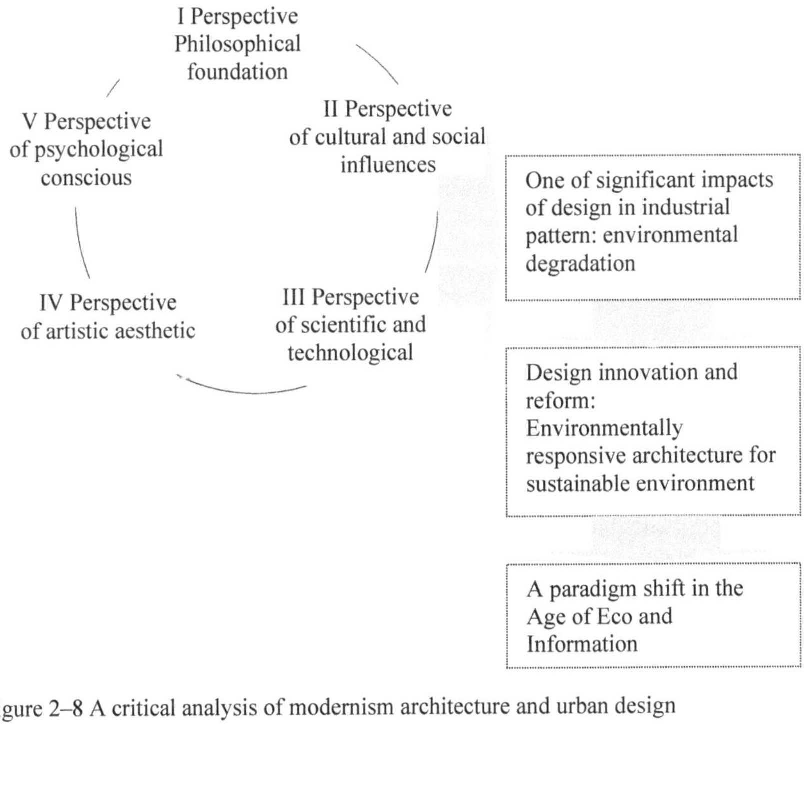

Figure 2–8 A critical analysis of modernism architecture and urban design

2.2 A new worldview of the Post-Modernism: from simplicity to complexity

As argued by Jencks (1997), the Western industrial society has been complained of losing its orientation, disintegrating into angry fragments for more than fifty years. It is generalised as cultural crisis in people's daily life, along with environmental crisis and economical crisis, which emerge especially remarkable in the early of 21st century. For example, modern science are criticised only producing partial truths, while traditional religion, as another principal means to explain ultimate truths, is also inadequate at interpreting and revealing the complexity of the universe. The diversity and complexity as the progressive consequence propelled by the industrialization and modernization is far beyond the capacity of the old paradigm of the modernism. In this backdrop, complexity science temporizes and emerges since the end of 20th century when modernization needs a new paradigm for ensuring the continuous progress in a sustainable manner, not only in environmental sense, but also in social-cultural-economic sense.

2.2.1 New worldview of the Post-Modernism: organised complexity

Historically, the 18th century science, following the Newtonian revolution, has been characterised as developing the sciences of organised simplicity, the 19th century science via statistical mechanics, as focusing on disorganised complexity, and 20th and 21st centuries science as confronting organised complexity. Specifically, traditional science in the model of machine since the 19th century tended to emphasize stability, order, uniformity, and equilibrium as the world view (Prigogine and Stengers, 1984; Nicolis and Prigogine, 1989). However, a paradigm shift which characterizes the accelerated social change nowadays discloses the aspects of reality of disorder, instability, diversity, disequilibrium, nonlinear relationships and a heightened sensitivity to the flows of time (Prigogine and Stengers, 1984; Nicolis and Prigogine, 1989). In addition, the complexity of physical-chemical and biological systems in nature since the 20th century has been disclosed, approached and modelled conceptually and mathematically in a new discipline of complexity science, to manifest a new worldview of evolution, instability and complexity of the universe (Prigogine and Stengers, 1984).

2.2.1.1 Complexity science to reveal the creativity of universe

The concept of complexity in natural science is deeply rooted in the laws of nature, applicable in biology, geophysics, material science, economics, sociology of animal and human populations, ecology, evolutionary theory, architecture, origins of life research, climate research, philosophy, art and even research for extraterrestrial life. Most of the real-world systems can be called complex systems, such as climate, living organisms, ecosystems, economic systems, hydrodynamic systems, biophysical socioeconomic systems, industrial systems and social system. These complex systems can not be guided like as linear mechanical devices as the modern science of simplicity in reductionism, mechanism, and linear determinism, but obeying non-linear laws of self-organisation, as the manifesto of the Post-Modernism (Jencks, 1997; Gleiniger and Vrachliotis, 2008).

In general, complexity is caused by the dynamics of interacting components within a system during self-organisation to adapt to the stimulus from its host environment, which consequently shape a complex network topology, and exhibiting observable structures, properties and behaviours (Ho and Saunders, 1979). The key themes of the causality of complexity are identified as attractors, topology of a system's phase space, and uncertainty regarding to the information of the systems. In biology, for example, the concept of complexity is associated with (Casti and Karlqvist, 1986; Nicolis and Prigogine, 1989; Kauffman, 1993; Adami, 2007):

- the population of living organic systems components,
- the connectivity of the population,
- information communication among them, and
- the behaviours of individual constituents of the population in systems;

- furthermore, evolution plays an important role in the complexity of observed behaviour in the view of their morphology and their functioning, including reliability, adaptation, growth and evolution under certain structural instructions;
- another complexity is physical complexity, referring to dynamic processes in evolutionary biology theory (Adami, 2007), i.e., the complexity increases after the evolution of self-organised system.

In other words, the studies of complexity are composed by conceptual and formal structures of system, the evolution process as a complex dynamics and the emergence of complexity after evolution (Casti and Karlqvist, 1986). Modelling complexity in nature is therefore to assess the nonlinear character of the underlying dynamics, to identify a set of variables capable of showing instabilities and bifurcations, and to find natural laws or mechanisms generating morphology and function.

The definitional authority of complexity science was located indisputably in the gravitational fields of cybernetics, computer science, information technology, quantum physics, fractals theory, chaos theory, catastrophe theory, emergence theory, nonlinear dynamics theory. The languages described in complexity science include non-equilibrium stability, bifurcation and symmetry breaking and long-range order, which involve the deterministic theory of dynamical systems (Hirsh and Smale 1974; Kauffman, 1993). To monitor and measure complexity, the definition of complexity in mathematics is given as the amount of information needed to describe a process, a system or an object, such as algorithmic complexity, and statistical complexity (Li and Vitányi, 2001; Prokopenko, Boschetti and Ryan, 2007). In addition, in complex system science, there are several approaches describing the complexity of a system, by structural complexity and functional complexity (Kay, 1984).

2.2.1.2 Evolution and self-organization of the universe for organised complexity

In contrast to the modernism cosmological viewpoint of gradual and continuous developments of the universe in mechanism, linearism, reductionism, the creativity of universe is now interpreted in the post-modernism complexity science as self-organising in feedback for the emergence of organised complexity; the evolution of the universe is interpreted as phase transitions in spontaneous change and quantum leap. Theoretically, the unpredicted and underdetermined aspects of the universe is interpreted in phase transitions of dissipative structures, i.e. the spontaneous change or the emergence of new organisations in a dissipative structure, due to external energy, heat and information, and the internal pressures as well (Nicolis and Prigogine, 1989). This cosmogenic orientation of nature going through sudden phase transitions in organisation discloses the fundamental principle of the universe in increasing complexity and higher levels of organisation, creative, open, dynamic, surprising in its active energy, its growths and sudden leaps, its beautiful twists, curls and turns, and its catastrophes (Jencks, 1997).

Such complexity and self-organisation of nature has been argued in "Gaia theory" in earth science (Lovelock, 2006), and cellular automata in biological science, for example, all living organisms, and even super-organisation, such as a beehive, city or any type of social organisation (von Neumann, 1966; Castrigiano and Hayes, 1993). Every being, from the atom to the nation to the galaxy, has a quality of self-organisation and a degree of freedom, the ability to reorganise itself through feedback. Each is balanced passion and reason in a delicate zone between too much order and too much chaos; this boundary condition is the place of maximum complexity and computability; everything in nature and culture is pushed toward this creative edge by evolutionary pressures, by natural selection and internal dynamics. However, the unpredictable, abrupt aspect of this new world-view also shows that violence and mass-extinction are essential parts of evolution.

2.2.2 Redefining complexity and creativity of architecture in the Post-Modernism

2.2.2.1 New worldview, new design strategies and new aesthetics

As a manifesto of the new world view in the post-modernism, complexity science claims the universe is unpredictable, dynamic, complex, self-organising in feedback; it rediscovers aesthetic and spiritual meanings of nature, which is denied by modernism (Jencks, 1997). With this inspiration, the fundamental intellectual and philosophical capacity of man and the human society is also meant to be self-organizing, in which the mind and culture are typical of their creativity, under the driving forces of human desire for learning, mankind evolving and progressing forward high level civilisation above ignorance. To reflect this new world view, the progress in science, the spirit of culture and the aesthetics of life, architecture and urban design will be able to departure from the modernism in a new expressive languages, conceptual, aesthetic, motivation, objective and philosophical aspects. Especially, confronting with the environmental crisis since the late 20th century, an alternative innovation in architecture and urban design from environmental perspective will be possibly realised.

2.2.2.2 Restoring complexity in post-modernism architecture and urban design

Theoretically, Venturi (1950) was one of the distinctive promoters of complexity in architecture. The principle concern for the re-establishment of complexity is to restore to both art and everyday life the rights deprived of by the strategies of simplifying systematization that governed the abstract model of architectural design prevailing in Modernism. The first stage of complexity in architecture is argued by Venturi (1988) as the collage of pre-existing, well-known solutions, the manipulation of classicism, Modernism, or any familiar ground. A recent argument on the complexity of architecture by Hagan (2001) is inclusive of topology, morphology, biology, geology and complexity currently swarming amongst the architectural intelligentsia. Another theoretic standpoint by Vrachliotis (2008) argues three lines of development on the complexity of architecture: a gestalt-psychological line, a cybernetic line and a biological-algorithmic line. In practices, the concept of complexity has been symbolically interpreted and presented in the work of leading architects, such as Peter Eisenman, Frank Gehry, Denial Libeskind (Jencks, 1997), Zaha Hadid (Pulselli and Tiezzi, 2009), along with ecological and organic designs. In urban design, with the fractal paradigm learning from fractal theory, planners can now get closer to the fine-grained, subtle growth of real cities and work with fractal design, which is in contrast to the mechanistic models that Le Corbusier and other Modernists imposed on this process of life.

2.2.3 Summary

Modernism architecture and urban design in the linear industrial pattern are developed in machine aesthetics of formalist, functionalist and mechanism, without environmental commitment and other social imperatives, as shown in Fig 2–9, which is especially challenged by the environmental crisis of climate change and global warming in the late 20th century. Accelerated by a series of revolutionary progresses in natural sciences, a new world-view of the Post-Modernism promotes a shift from simplicity of Modernism to organised complexity. With this new world view, the complexity of sustainable design will be possibly resolved by expanding the inadequacy of modernism design agenda with trans-disciplinary knowledge and a systematic approach (Birkeland, 2002), ranging from the scales of megalopolis to the choice of construction systems.

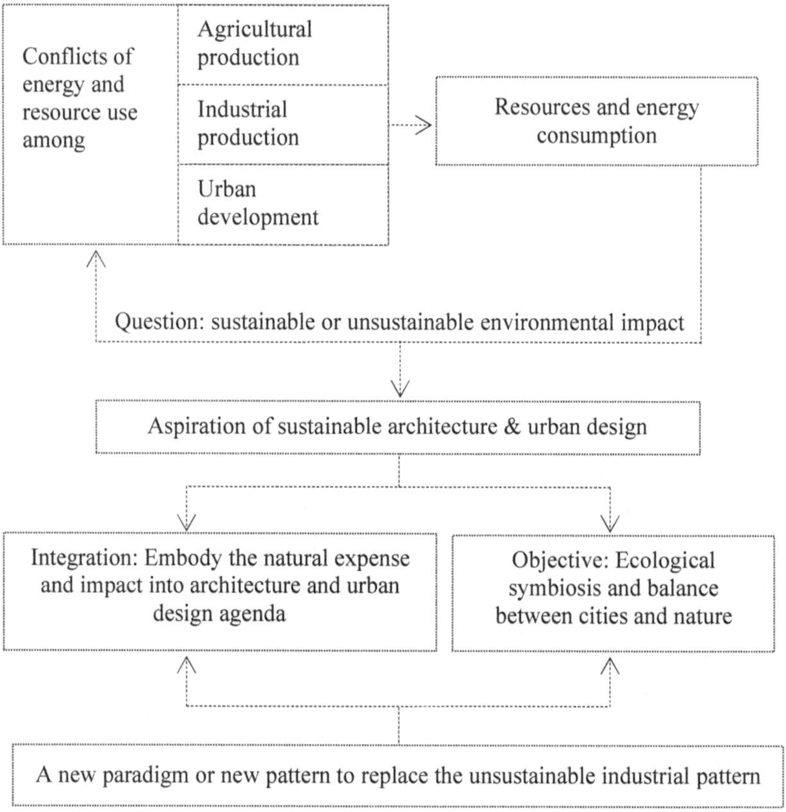

Figure 2–9 Analysis of environmental aspects of architecture and urban design

2.3 Contemporary campaign for sustainable design

Confronted with the contemporary environmental crisis, a necessity in high priority is to reinforce the environmental relevance of architecture and urban design towards a sustainable built environment, so that buildings and cities would be earth-friendly to alleviate environmental problems and to restore the natural ecosystem. This movement of sustainable design will liberate architecture and urban design from the constraints of existing industrial technology, politics, and economics, changing the profession's basic philosophical and aesthetic values, design pattern and methods (Ray-Jones, 2000; Wines, 2000). Sustainable design cannot be viewed as technical problems only; rather, it is complex hybrids in response to the challenge of environmental sustainability which is shaped by the widely differing motivations and competing social commitments.

2.3.1 Environmental awareness in the building industry

Since the energy crisis in the 1970s, being warned of oil shortages, embargoes and war, along with heightened concerns over pollution, environmental degradation, and resource depletion, the environmental impact of architecture and urban design has widely been criticised and questioned, with an emphasis to improve physical performance and efficiently use energy and resources in buildings and cities (Sustainable architecture white papers, 2000; Guy and Moore, 2005). The severity of the industrial pattern in modern design has disvalue the relevance of buildings and cities to the contemporary age, while the original visual factor becomes marginalized (Portoghesi, 2000; Hagan, 2001).

To explore alternative propositions of sustainable design, especially enhancing buildings and cities' environmental performances and reducing their unsustainable environmental impacts, questions on the building industry are addressed from the perspectives of ethical concern, design motivation, principle, concept and technique, as follows:

- Question on the sources of energy and resources use: is it possible to adopt the mechanics of buildings in such a way by using other energy, i.e. biomass and solar energy and resources, to replace fossil fuel, for less, neutral, and even positive environmental impact?
- Question on the pattern of energy and resources use: how to improve the linear metabolism of energy and resource for buildings and cities in an alternatively efficient manner towards sustainable environmental performance?

- Question on technologies on environmental management: to facilitate environmental management in the built environment, which technologies and techniques can be devised to effectively indicate, monitor, assess, adapt and optimize the environmental performance of buildings and cities over the life-cycles?
- Question on design configuration: how to configure environmental constraints, i.e. ecological dependence and environmental impacts, as essential design criteria for environmentally responsive buildings and cities?
- Question on ethics (Hagan, 2001): why human habitant by buildings is designed in such industrial mechanics, intervening into nature by consuming so much energy and resource and producing such great amount of negative environmental impacts?

2.3.2 Motivation and commitment of sustainable design

In contrast to the loss of a connection to nature caused by design in the industrial pattern, the values of natural capitals, i.e. healthy air, pure water, efficient ozone layer, clean sea, fertile land and great species diversity, are now appreciated in the vision of in harmony of buildings and cities with the natural environment, which has arisen from the nationalistic, ideological, religious or economic groups (Wines, 2000). Furthermore, an integrated agenda of inter-disciplinary research on sustainable design has been engaged by scientists and sociologists, architects and engineers, philosophers of environmental ethics as well as citizens themselves. Such integrity brings the needs and perceptions of social groups into the process of developing sustainable buildings and cities. Thus, design problems are reframed and embedded into a broader picture of socio-technical system change (Guy and Moore, 2005). The motivation, objective, philosophy of sustainable design are generalised as follows:

- Motivation: to advance the purely self-interest motivation of human survival (Milton, 1996; Wines, 2000);
- Objective: to build in ecological principles for sustainable symbiosis of man-made space with nature (Guy and Moore, 2005);

- Philosophy: to solve conflicts surrounding the issue of environmental crisis in resistance or in adaption (Wines, 2000).

In brief, neither the environmental protection nor the social interest can be given paramount status, but it is essential to explore a process of co-construction for a harmony between them. As Schumacher (1973) notes, an industrial system could be called efficient only if it has obtained strikingly successful results in terms of human happiness, well-being, culture, peace and harmony. Thus, the motivation of sustainable design in buildings and cities is ideologically embedded in contemporary social and culture imperatives and values; it will adapt some strategies and reject others in terms of the social structures and institutional values (Wines, 2000).

2.3.3 Institutionalised definition of sustainable design

Three years after Brundtland Committee's definition of sustainable development as a compromise in the growth and no-growth with an emphasis of limited carrying capacity of the planet, the Commission of the European Communities released a far more holistic explanation of the reasons behind accelerated environmental degradation. Its Green Paper on the Urban Environment (1990) marks the beginning of environmental policy-making in urban areas within the European Union, which contains specific prescriptions for planning and building strategies to improving environmental sensibility. In 1996, the second UN conference on human settlements (Habitat II) in Istanbul put forward ways of applying sustainable principles in buildings. Another international thinking on environmental design can be seen in the green design handbook "The Green Vitruvius", UIA, "Declaration of Interdependence for sustainable future", Green Building Challenge (Gauzin-Müller and Favet, 2002). In brief, from the 1990s, new sources of form and content to transform buildings more energy-efficient are considered with significance in energy policy.

Specifically, UK's Building Services Research and Information Association (BSRIA) define sustainable construction as the creation and management of healthy buildings based upon energy and resource efficiency, and in ecological principle. The American Institute of Architects has adopted ecology as an official part of its programme, and a World Congress of Architects has signed a Declaration of Interdependence concerning the environment (Jencks, 1997).

2.3.4 Characteristic features of sustainable design

In terms of Brundtland's definition of sustainability, to meets the needs of the present without compromising the ability of future generations to meet their own needs, sustainability is evocative of optimistic and protective ideas, recalling good common sense, fully engaging in the world in a way that is empathetic, intuitive and aesthetic, in a critical response to the imperatives of ecological, social and economic forces of the late Industrial Age (Wines, 2000b). Hence, sustainable architecture is used to describe environmentally responsive architecture, the scope of which consists of efficient energy and resource use, human healthy, ecological principles, economically viability and local contextual sensitivity, through technology innovations, traditional wisdom and institutional change for a global responsibility (Guy and Moore, 2005). In brief, the prototype of sustainable design is in new aesthetic values and an integration of nature, technology and humanity, re-imaging the relationship between human beings and the natural ecosystem systems (Sustainable architecture white papers, 2000).

The philosophical underpinning of modernism design, anthropocentrism, which induces human habitual isolated and detached from the ecological awareness, is exchanged for the reawakening of "oneness with nature" (Wines, 2000b). Sensitivity to nature has been passionately advocated by landscape architect Ian McHarg (2006), who argued no human intervention into nature should be considered without a thorough analysis of the compatibility between a building and the ecosystem, including the specific topographical and microclimatic conditions of the site (Revkin, 2001; Steele, 2005).

It is now becoming more widely acknowledged that sustainability is a contested concept open to diverse interpretations and design strategies by utilising relative technologies. The most popular view of sustainable design is roughly synonymous with energy and resources efficiency (Steele, 2005). In either an active or passive mode, buildings and cities are designed climatically and culturally effective over time, to respond to local typologies, regional microclimates and materials, (Guy and Moore, 2005).

2.3.5 Technological implementations of sustainable design

Since the Industrial Revolution, when technology has been unleashed by the socioeconomic forces, profoundly affected the cultural, political and psychological conditions of industrial societies, technological determinism believes that technology could solve any problem and predicates that social benefit is entirely dependent on scientific discovery and technological innovation (Steele, 2005; Guy and Moore, 2005). These technological campaigns, looking to science for sustainable environmental management and integration in architecture, cities, energy systems, transport and industry, began in the 1920s with the work of architect Buckminster Fuller, moving through the 1960s and 1970s with the emergence of a new High-Tech approach in science-led aesthetics (Steele, 2005).

2.3.5.1 Sustainable aesthetics by technological innovations

In the faith in modern science and technology, hybrid technologies utilising advanced technology and traditional systems have been developed in sustainable design in the principle of contextually responsive and functionally viable. It expresses in new aesthetic languages of its environmental and social commitments, shifting from the previous modernism's abstract form and function (Steele, 1997; Guy and Moore, 2005). These sustainable aesthetic technologies are generalised as follows:

– Passive environmental design (Sustainable architecture white papers, 2000; Hagan, 2001; Langston and Ding, 2001): passive solar design principles of the orientation of windows related to the position of the sun, thermal mass, natural ventilation, shading and sunlit areas, to reduce heating and cooling requirements along with day-lighting and energy-efficient equipment and systems; in

warm climates, strategies in excluding the heat of the sun and harnessing the cooling effect of breeze; to reduce surface to volume ratio (SVR) to improve the heating and cooling efficiency of a building, by minimizing the exterior wall and roof areas to produce a more compact buildings shape;
- Low-embodied energy strategy: reducing the energy intensity of buildings through techniques by insulting materials, low-energy lighting and natural ventilation, wind generators, increased levels of thermal insulation by thermal mass for passive heating and cooling, air-to-air heat exchangers, heat-recovery ventilation systems;
- Ecologically sound building materials, e.g. façade design for environmental sensitivity for healthier interiors and better glazing (Hagan, 2001; Margolis and Robinson, 2007);
- Renewable energy uses: solar collectors, photovoltaics and other renewable energy resources to replace fossil fuel used in buildings to reduce emission and waste;
- Sophisticated techniques for measuring the effectiveness of various land use configurations in urban areas for the microclimate control and "urban heat-island" (Steele, 1997).

2.3.5.2 Sustainable assessment of environmental performance

Environmental performance of buildings can effectively be predefined or assessed through the use of objective technical analysis such as life-cycle analysis, ecological footprints analysis or environmental assessment methods (Steele, 1997). These environmental evaluation methods have tended to be portrayed as objective, value-free tools to interpret and assess sustainable buildings on the basis of their environmental attributes such as energy efficiency. However, these methods of integrate quantitative building performance data are mostly post-occupancy analysis of people living in buildings and cities in the direction of energy efficiency.

2.3.6 *Challenges of sustainable design*

Embedded with social, economical and environmental imperatives, progressed with scientific and technological innovation, sustainable design effectively expresses the societal sub-consciousness in new aesthetics. However, learning from human civilization over the last centuries, any paradigm shift may be hindered by psychological and cultural commitments of the societies (Wines, 2000), and methodological and technical limitations and constraints.

2.3.6.1 *Philosophical paradigm shift in concept and aesthetics*

Idealistically, the construction of sustainability lies in encompassing ethics, economics, sociology, history, ecology and biology for a start, a wide variety of information to be assimilated and converted into a final design solution with unique aesthetics of sustainable architecture (Steele, 1997). Although the environmental awareness of global climate change emphasizes environmental responsibility in sustainable architecture, in thirty years of sustainable design, individual architects have achieved relative little. Environmental commitments in passive environmental design have been mostly focused on technical repair, not yet based on metaphorical references of cosmological philosophy to reveal human associations with the earth (Jencks, 1997). More advanced design approaches looking beyond solar panels and photovoltaic cells are expected as a fundamental revolution in sustainable design. In brief, the key challenge to the art of sustainable design is to integrate conservation technology with an earth-centric philosophy, to construct the human habitat with sensitivity to the sustainable environment principle and translate this message into a new architectural iconography, feasible as both local interpretation and global implementation.

2.3.6.2 *Methodological backbone of science for sustainable design*

Since the Renaissance, architecture is a combination of art, science and technology (Vallero and Brasier, 2008). The rationalisation from concept to completion was processed in a desire for an increasingly rich aesthetic expression, contributing to the growing complexity of architectural treatises. However, the basic thermodynamics as a fundamental methodology for studying environmental aspects of buildings and cities has been ignored, leading to the deletion of their environmental consciousness. For example, by the 1990s, Edward Mazria (1997) argues most contemporary buildings, like machines, are provided with the energy they need to run, and do so according to the Second Law of Thermodynamics: importing and converting energy to run equipment and lighting fixtures, and dissipating the imported energy as waste heat. Without an uninterrupted supply of imported energy; in result, such isolated systems become uninhabitable, which are intentionally isolated against the environment for as long as possible to preserve artificial internal conditions. In contrast to this conventional scheme, based on the empirical experience Mazria has formulated a theory for environmentally responsive architecture, mimicking nature more closely, which would operate as an open rather than an isolated system.

In addition to this scientific proposition based on the Second Law of Thermodynamics, ecological design, or biological, green or Gaia architecture has been variously termed to sustainable architecture (Guy and Moore, 2005). The ecological definition expresses ecological values of architecture in terms of nature science of ecology, which opens a paradigm shift from machine in the Industrial Age to nature in the Ecology Age. One of scientific propositions on the revolution on modern architecture and urban design towards sustainable environment is systematically summarized in Tab 2–2:

Table 2–2 A revolutionary design approach in science to sustainability

Phenomena	Physical reality of environmental crisis: un-sustainability
Background	Urban growth & economic development in industrial culture
Evaluation	Quantitative approach of ecological footprint to reveal the reality of environmental crisis
Problem	Ecological system analysis of the crisis and challenge for sustainability
Framework	Emphasis of the ecological relationships of human habitat, i.e. urbanization human region and hinterland, based on theoretical foundation of thermodynamic laws
Proposal	A new design method for sustainability in complex systems science

In summary, one of present insufficiencies of sustainable design is mostly due to a missing of scientific methodology and framework into design configuration, which would rationally structure the entire set of environmental interrelations between buildings and cities with the natural ecosystems (Steele, 1997; Guy and Moore, 2005; Vallero and Brasier, 2008; Steadman, 2008), as shown in Fig. 2–10.

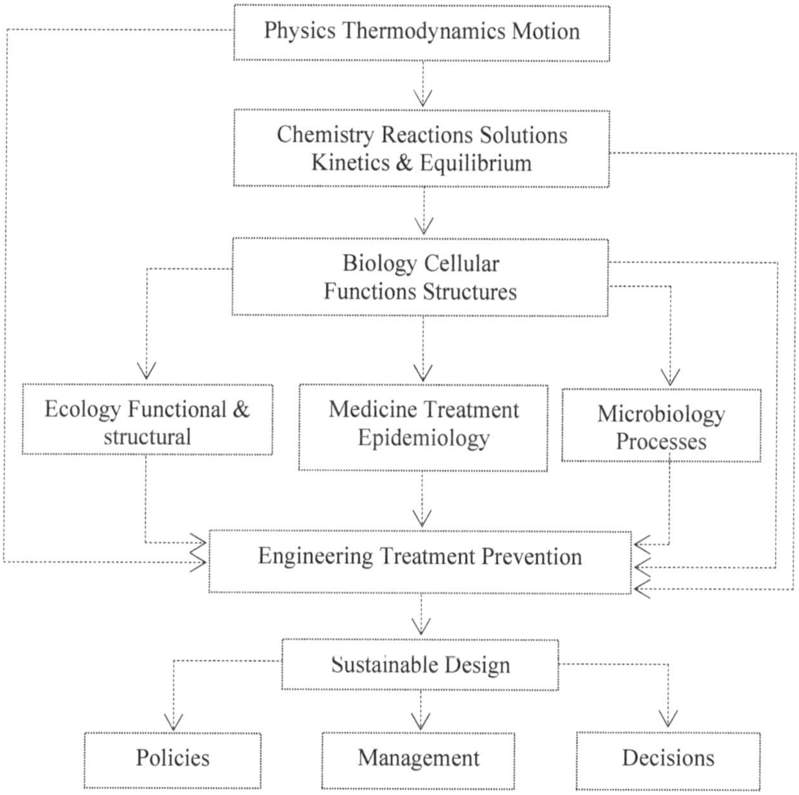

Figure 2–10 Cascade of science leading to sustainable decision

2.4 A paradigm of nature for sustainable design

Nature is primary, metamorphic, and endlessly ambiguous, rich in associations, which offers a universal source of ideas and symbolism, and a more relevant visual content imagery in the art of architecture (Mclennan, 2000; Wines, 2000b). The tradition respecting nature as an inspiring source for creativity and in a ritual admiration for harmony between man and nature can be traced back to naturism in the Pre-Industrial Age. More recently, the profusion of evocative subject matter associated with nature consists of its comfortable scale relationships to the human body, the complexity of its structure, and the science of describing its dynamic processes. Through fundamental human perceptions and modern progress in ecology and other sciences, nature has been appreciated as methodological paradigms and inspirations, such as the morphology of form, structure and function in organic architecture and urban design, the dynamic mechanisms of self-organisation, optimisation in evolutionary and generative architecture and urban design, spiritually symbolic value of harmony between man and nature in vernacular architecture. All of these appreciations on the paradigm of nature provide environmentally-responsive design approaches for sustainable buildings and cities.

2.4.1 Ideological interpretations of nature in religion and philosophy

Nostalgia to the Pre-Industrial Age, including aboriginal and agricultural societies which were removed from the modern age, is a gesture in responses to the environmental damage induced by drawing unrestrainedly on the earth's surface as the wasteful legacy of industrialized societies is. In these civilizations, the philosophy of sustainability originated is associated with the societal perception of eternity, in cosmology of eco-centrism or naturism in contrast to anthropocentrism in the industrial societies. It suggests a paradigm shift from machine to nature, to reconnect the contact of man and nature, to regain iconographic and functional relevance of human habitat (Portoghesi, 2000; Wines, 2000; Birkeland, 2002).

Especially, the philosophical foundations of nature, which inspire eco-centrism, naturism, bio-regionalists, Gaians, eco-feminists, eco-Marxists, bio-centrecists, eco-anarchists, deep ecologists and social ecologists in the modern age (Guy and Moore, 2005), can be traced back to the nature-oriented religions of Zen, Buddhism and Taoism in the East, indigenous culture of the aboriginal societies (Wines, 2000), as well as environmental friendly philosophers like St Francis[13], Spinoza[14], Gandhi[15] and Carson (1963) in the West.

In contrast to anthropocentrism, the philosophy of Taoism[16] emphasises the priority of humanity's connections to nature in the East Asia two thousand years ago. Its ideology is generalised as "Tiān Rěn Hě Yī" (Jin, 2005) in Chinese, which means "oneness with nature",

---

13 Philosophy of non-violent action, Saint Francis (1181–1228) of Assisi brings to mind an "animal lover". He is always depicted with a bird on his shoulder. He is a powerful reminder that Faith in a Loving God must be at the core of anyone hoping to practice Non-violent Action.

14 Baruch Spinoza (1632–1677) is Dutch philosopher of Portuguese Jewish origin. Revealing considerable scientific aptitude, the breadth and importance of Spinoza's work was not fully realized until years after his death. Today, he is considered one of the great rationalists of 17th century philosophy, laying the groundwork for the 18th century Enlightenment and modern biblical criticism.

15 Mahatma Gandhi (1869–1948): "God forbid that India should ever take to industrialism after the manner of the west." "If (our nation) took to similar economic exploitation, it would strip the world bare life locusts." He was the pre-eminent political and spiritual leader of India during the Indian independence movement. He was the pioneer of mass civil disobedience, firmly founded upon total non-violence, which led India to independence and has inspired movements for civil rights and freedom across the world.

16 Taoism refers to a variety of related philosophical and religious traditions and concepts that have influenced East Asia for over two millennia and the West for over two centuries. From the 3rd century B.C, it is based on a belief that all living forms can be resurrected on a higher level, to seek simplicity in accord with nature and reject worldly pursuits in preparation for a possible deification through infinite connectedness. The objective was to live beyond convention moral and social standards, believing that such limitations alienated one from nature. All the routine activities and constructions of civilization were considered to be of a lower order.

"man and nature in one" and "the harmony between the heaven and the human". It calls for man and nature in an ecological concordance for harmonious coexistence with nature, which was resolved through balancing the confronted conflicts with nature rather than aggressively assaulting nature. In brief, it reflects the agricultural civilization, when tangible knowledge is not fully developed to strengthen people with the idea of conquering nature; on the contrary, the ideological dependence on the natural environment was rather developed, which provided people with the comprehension of nature more powerful than man.

Based on the primary comprehension of man's symbiosis with and resilience on the natural environment in these agricultural societies, it is believed that the practice of human living should be a reflection of nature's infinite harmony, and the structure of the mind as an extension of all the components of nature and the universe. The cosmological concept has been frequently reflected in art works, as shown in Fig 2–12, a modern art work of landscape in man's mind. It embraced anthropomorphism and a view of earth's treasures as the mirror of an even greater paradise beyond terrestrial life. In summary, the cosmology of Taoism and Zen is earth-centrism.

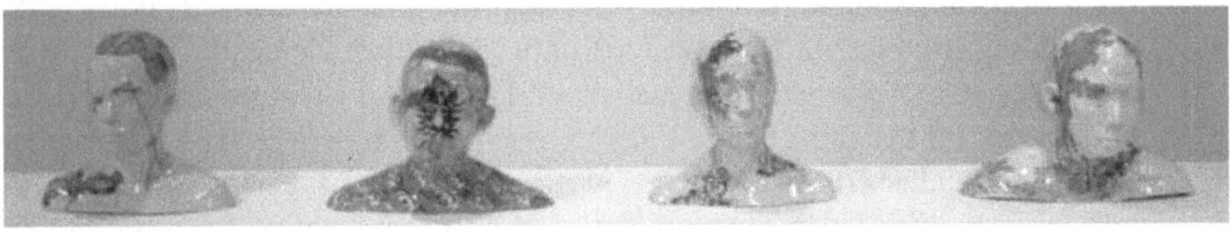

Figure 2–11 Sculpture by China/Australia artist Ah Xian[17]

One of philosophers on eco-centrism in the 20th century is Naess (1989), who radically advocated the deep ecology movement, the major purpose of which is to change the paradigms which have distorted contemporary society's value systems, economics and relationships to nature. It is claimed that human race does no more deserve of respect, protection, or continuity than any other organism. Instead, the phi-

17 Photograph from art exhibition at Modern Art Gallery of Queensland, Australia, by the author. Copyright permission for the reproduction in this book is granted by the Gallery.

losophy of eco-centrism or eco-philosophy argues that it is only through a massive reversal of techno-centrism and anthropocentrism that humanity has any hope of survival, and man should return to a humble role as only one small component of the total eco-system, to revert the industrial damage upon nature, not only from the standpoint of its ethical, but also of ecological principles. By this humbling, it is possible to realise human cooperation towards ecological harmony, rather than industrial aggressive progression any further (Portoghesi, 2000).

2.4.2 Scientific interpretations of nature in ecology and biology

In addition to the philosophical appreciation of nature into the modern world, it is recognised scientific interpretations of nature would provide more rational strategies to replace the entrenched industrial design pattern in an exploitation of nature. For example, ecology is a science studying the relationship between all living organisms and their natural surroundings (Peterson, 2001); biological science studies the individual living organisms in the natural eco-systems. Both of these modern sciences have inspired (Steadman, 2008) the conceptual and methodological paradigm for art and architectural creations.

Macroscopically, nature as a living ecosystem exhibits sustainable prosperity of vitality, diversity and complexity (Benyus, 2002). It is consolidated by microscopic interactions between constitutional biological organisms in the natural ecosystems and those interactions between biological organisms and the ecosystem, in evolutionary and generative manner by which feedback facilitates self-organising and self-regulating of natural ecosystem which has been studied in cybernetics (Wiener, 1997), and in morphogenetic fields (Arnheim, 1977). In the cycles of nature or the biosphere, materials are operated according to a system of nutrients so that there is no such thing as waste (McDonough and Braungart, 2002). This metabolism of nature is streamlining, material cycling and minimizing the use of raw resources and energy (Bisch, 2001).

In summary, in the 20th century, science underwent a major paradigm shift, from the projection of nature-as-clockwork to a new projection of nature as relative, dynamic and uncertain (Prigogine and Stengers, 1984; Wines, 1987). It is the flexibility, the reactive and transformational capacities of the ecological natural world that scientists are now seeking to emulate in materials, computers and robotics, rather than the traditional idea of the organics as forms found in nature, or as the use of natural materials, or as a perfectly organized whole in which all the parts are essential and proportionally related. It suggests a new rational reappraisal of nature to reveal the depth relationship between man and nature.

2.4.3 Morphological imitations of nature in organic design

Starting from the Renaissance, design in architecture imitates the mathematical proportional laws found in nature, e.g. the sets of harmonic relations, the relation of the parts to the whole (Hagan, 2001). In the modern time, this classical perception of nature is termed as organic, which means intrinsic entity in the philosophical sense, wherever the whole is to the part as the part is to the whole, and where the nature of materials, the nature of the purpose, and the nature of the entire performance become clear as a necessity (Wines, 2000; Steadman, 2008).

In this method, organic architecture morphologically mimics nature's patterns for the organic form, the compositional laws of organic form, and the logic relation of function and form. Specifically, the biological forms of plants and animals are recognised as the compositional laws and rules for form, function, structure (Vrachliotis, 2008) and materials in an aesthetic sense (Steadman, 2008), and as the methodological rules for the relationship between form and function in a logic sense. These morphological methods have been applied for the geometrical organization of the parts and structure, the topological relations, the structure of circulation routes in both buildings and cities (Portoghesi, 2000).

In the early 20th century, two of the greatest observers of natural phenomena were Gaudi (Bergós, 1999) and Wright (Wines, 2000). They studied how plants grow, analyzed land surfaces, evaluated seasonal change, researched geology, and learned from the most subtle levels of earth-centered cause and effect, the interconnected processes through which nature achieve its miraculous states of evolution and perpetuity. Gaudi's greatness of naturalism is his prolific invention of architectural forms, the unusual structural contrivances and the imaginative use of materials to express a unified sense of decoration, which characteristic of three traditional attributes of his work. For Wright, organic architecture is in the principles of being integrated with its context, from within outward with the form being converted as an expression from the inside philosophy and as the extension of natural surrounding (Portoghesi, 2000; Steadman, 2008). He understood complexity of nature carefully, often citing nature for its authentic and sincere forms as well as its ability to acclaim the gifts of matter by turning truth, sincerity and appropriateness into real dogmas (Pfeiffer, 1939; Steele, 2005).

2.4.4 Dynamic mechanisms of nature in evolutionary design

Another imitation of nature in design, particularly since the late 20th century, is the dynamic process of self-organization and evolution, generation, optimization, adaptation and other cybernetic mechanisms in nature (Portoghesi, 2000; Vrachliotis, 2008). All of these new natural patterns have been generalized into the new world-view of the Post-Modernism. From this new world-view, designers derive models and methods for evolutionary and generative design (Steele, 2005), complexity architecture (Ilagan, 2001), cybernetic architecture (Johansen, 2002), and "biomorphic" architecture through programming techniques of genetic algorithms (Steadman, 2008), which form the genre of Post-Modernism architecture and urban design. One of pioneers reflecting this new world-view of dynamic spontaneous self-organization is architect Frazer (Frazer, 1995), who develops how the information is encoded in the generation of architectural form by imitating the non-linear model in nature, the dynamic self-organizing activity characteristic of all non-linear phenomena in nature. At the urban scale, Portoghesi (2000) argues the spontaneous growth of cities in the analogy of organism in nature: cities are alive and functional as any living creature, which follows the same process of miniaturisation and complexity to become a lively container for the social, cultural and spiritual evolution of human kind.

2.4.5 Environmentally-responsive design in vernacular architecture

In addition to morphologies of organic architecture and mechanics of evolutionary and generative architecture, nature also inspires environmentally-responsive design philosophy, which is typically embodied in vernacular architecture. The quest for harmonious balance between man and the natural surrounding has been adopted out of necessity into vernacular architecture and urban design, which is argued (Hagan, 2001) as a primary prototype of environmentally responsive design. Its practical earth-centric philosophy is to mitigate the environmental deficiency in modernism architecture and urban design in the Industrial Age, by taking advantage of the differentiations and gradients available in local climates and topographies, by passive demanding and consuming energy, low-technologies of structure, local material use, space-and-form-learning from nature, and construction technology developed out of regionally accessible materials. In result, buildings do not to press unreasonable stress on the natural environment, not to negatively affect regional ecology, and not to require high levels of energy consumption for heating and cooling.

Thus, vernacular architecture is environmentally responsive to the demands of climate, topography, and agriculture as the principle means of survival (Wines, 2000). In aesthetics, vernacular architecture is the extension of their natural contexts, not only represented in aesthetic forms, but also favouring physical interactions with the natural environment, the psychological relation of people to all of the various components of a building, and expressing spiritual symbol of harmony between nature and man. Examples of vernacular architecture and urban design can be found in ecologically balanced Mediterranean city (Salat, 2006), the fully environmentally integrated traditional Chinese, the combination of natural elements in Japanese architecture, and hand-built dwelling in Australian aboriginal habitats (Wines, 2000).

As a manifestation of this cosmology and philosophies of Taoism and Zen, traditional Chinese architecture and landscapes are integrated passively to nature in a manner of naïve and unsophisticated naturalism. It formed the art of architecture and landscape as being nature, not to form a purely observational perspective. Buildings are seamlessly merging with the natural landscape in a reverent way. In the design agenda, the theme of "borrowed scenery" combined with a concept of "inside-and-outside" is responsible for some of the most profound relationships between buildings and their natural sites. These sceno-graphic fantasies exhibited the harmonious integration of building and landscape, and reflect the organic and asymmetrical characteristics of a perceived paradise, nature. In theory, Chinese geomancy of Feng-Shui is a specific knowledge that tends to guarantee harmony and continuity between architecture and nature (Portoghesi, 2000). China's traditional urban morphology was characterized by passive use of natural resources and low environmental impact.

Another rich legacy for the practice of sustainable architecture is traditional design strategies in Japanese architecture. Its underlying principle is amity with the environment in the logic of natural growth (Gleiniger and Vrachliotis, 2008), the strategies is being close to nature and tailored to the local climate and terrain, and the spatial form brings the effect in spaces that harness the benefits of nature in ways that provide comfort with minimal energy consumption and load (Ray-Jones, 2000).

However, the aesthetics of vernacular architecture, although rooted in religious and civic celebration, was simply a result of lack of fossil fuel-driven machinery, the unavailability of industrially manufactured building products, and the inherent appeal of hand-made architecture. Furthermore, the philosophy, method, pattern and form of environmentally instructive architecture and urban design developed in the Pre-Industrial, agricultural and aboriginal societies are challenged by the complex demands of contemporary urbanisation and industrialisation, including complex modern urban infrastructure in the industrial societies, its alarming population explosion and universal dependence on fossil fuels and international communications (Ray-Jones, 2000; Hagan, 2001).

2.4.6 *Summary*

To realise sustainable design in buildings and cities, it needs a paradigm shift from machine in the industrial age to nature in the age of environmental crisis, which affects conceptual identification, technical implementations, aesthetic expression and philosophical underpinning in architecture and urban design. In the Age of Ecology and Information since the late 20th century, sustainable design will be explored more concretely with methodological frameworks from natural sciences, such as ecology, biology, thermodynamic science and information science.

2.5 Ecological design for sustainable environment

Ecological design is one of traditional treatise of nature-inspired design in architecture. Theoretically, ecological design is proposed that buildings and cities actively interact with nature in the model of ecological entities, taking use of energy and natural resources from nature as inputs, emitting pollution as outputs, adapting and optimising their environmental performances in nature's non-linear feedback mechanism to replace the unsustainable linear industrial pattern. However, in contemporary design practices, ecological design's objectives, strategies, techniques, evaluations, are still under-developing due to certain conceptual and technical barriers and constrains yet. In such a backdrop, particularly since the mid-20th century, digital design techniques have been developed for efficiently modelling and simulating the complex environmental performance of buildings and cities towards sustainable environment.

2.5.1 Theoretic explorations of ecological design

2.5.1.1 Aims and principles of ecological design

Initially, ecological design is established in an aim to acknowledge, facilitate and preserve the interrelationship of nature and man-made environment for an ecological balance; in the methods by learning and applying the principles of nature for buildings and cities. A primary definition of ecological design by Ryn and Cowan (1996) is to transform matter and energy using processes in the manners which are compatible and synergistic with the nature ecosystems; to directly incorporate the natural ecosystem into buildings and cities improving the functioning of the built environment; in addition, human-nature interface is managed for the benefit of both buildings and the natural ecosystem. Another similar proposition for ecological design is proposed by Yeang (1995), a pioneer in ecological design since 1970s, generalised in Table 2–3, followed by a proposition of ecological urban design, generalised in Table 2–4.

Table 2–3 A prototype modelling of ecological design in buildings

Index	Content
Subjective	Design subjective is the interaction of building with the ecosystem, building is integrated with its environmental context, design be responsive and anticipatory to result in beneficial effects for natural systems.
Perspective	Design launches from the ecosphere perspective in a closed system, matter, energy, the biosphere's limited productive and assimilative capacity for human-induced consumption and waste to be considered into design to conserve ecosystems.
Impact	The impact of the design be considered over its entire life cycle, focus on matter-energy life-cycle impacts
Approach	Design will be approached holistically, using integrated thinking approach rather than in a fragmented manner, because of the complex impacts of built environments on nature.

Table 2–4 Key indexes of ecological urban design

Indexes	Notions	Definition
Sustainable vision	Ecological balance	Using scarce natural resources in a way that safeguards the interests of future generations. Natural resources need to be preserved and treated as an ecosystem rather than as a commodity.
Sustainable evaluation	Ecological footprint (Rees and Wackernagel, 1996)	The "ecological footprint" and the "life-cycle assessment" of environmental impact of building, which translate the amount of resources swallowed up by cities/buildings and the amount of wastes they produce into areas of productive natural ecosphere.
	Sustainable criteria of ecological footprint	The consumption of renewable and replenishable energy and resources must not exceed their production in nature; the production of degraded energy and matter by cities must not exceed the assimilative capacity of local ecosystems or the ecosphere.
Sustainable mechanism	Autonomous ecological system	Cities should be apprehended in their entirety as closed-loop ecological systems, as autonomous as possible by optimizing resources reuse, with input and output levels (resources and wastes) that respect the carrying capacity of the earth, with limited productive and assimilative capacity of ecosystems for human-induced consumption and waste.
Sustainable Approach	Circular and closed loop "metabolism"	Consumption is reduced and the reuse of resources is optimized.
Definition of ecological cities	Complex ecological integration of natural and man	Urban areas are defined as the physical environment that consists of "a complex mix of natural elements including air, water, land, climate, flora and fauna and the built envi-

		ronment that is constructed or modified for human habitation and activities, encompassing buildings, infrastructures and urban open spaces" (Rees and Wackernagel, 1996)
Sustainable Goal	Ecological environment quality (Hardy et al., 2001) Provide the environmental conditions they can use the health of urban citizens and reduce the vulnerability of the population, including basic infrastructures and services such as adequate provision of water, sanitation, rubbish collection and drainage for all the urban area and its citizens. (function of nature to city) Provide citizens with high quality urban environment that protects the natural and cultural heritage and provide outdoor comfort and the necessary urban space for the well being of the city dwellers (mixed functions of city) Reduce the risk of chemical and physical hazards in the everyday life of the city (healthy city) Reduce as much as possible the shift of the environmental load and cost generated by cities to the inhabitants and ecosystems surrounding the urban area. (city's consumption of nature) Ensure that the consumption of resources and goods and the corresponding generation of matter and degraded energy are compatible with the limits of the material capital and do not transfer environmental load and cost to future generations or to other human groups. (limits of nature capacity)	

To realise these aims and principles, several ecological design approaches have been explored contemporarily, such as positive design by positive contribution to nature, generalised as follows.

2.5.1.2 Restorative design for positive contributions

A conceptual proposition of positive development in the built environment is articulated by Birkeland (2008), in which positive contributions of buildings and cities to the natural ecosystem and the sociological system are emphasised. Positive development is defined as reversible development which achieves net positive impacts over the life-cycle by increasing economic, social and ecological capital, health and resilience in the built environment. It aims to enhance and restore the environment without sacrificing resources, not only delivering clean energy, water, air and soil, but also increasing carrying capacity and ecosystem services, going beyond remediation to generate net positive impacts and increase ecological productivity and resilience of the environment. Thus, in the proposition of positive development, buildings and cities would work ecological productive systems, infrastructures for the natural eco-systems. To realise this conceptual proposition of positive design, a technical proposition of "resource autonomy" supports with a microscopic mechanism of energy and resource use in buildings and cities for positive contributions.

2.5.1.3 Resource-autonomy for positive dependences on nature

The theoretical autonomy of architecture can be traced to Vitruvius (Ingersoll, 1996), whose treatise became the basis of the Western canon since the 15th century. Lately, Leon Battista Alberti heightened the autonomy of architecture by the avoidance of natural metaphors, i.e. the building appears a single, integral, and well-composed body, rather than a collection of extraneous parts. In the modern age, the autonomy is specially referred into the energy and resources consumption in buildings.

In 1970s, an autonomous-house concept was proposed by Pike at Cambridge, UK, to devise a servicing system for houses reducing dependence on limited localised consumables (Pike, 1974). This idea is taken further by Lapthorne (1973) with specific design agenda, including relevance to the natural climate and in unique visual design languages. In 1975, a concept of autonomous house is re-defined by Vale and Vale (2000), house operating independently of any input except those of its immediate environment, which has been applied by an autonomous energy school in Hungary (Imre, 1991), i.e. use photovoltaic, solar collectors and wind generators to meet its electricity and hot-water demands, with a bio-briquette-fired boiler for back-up heating, collecting water from a bored well and create biogas from sewage. Similarly, in a guide to the solar houses of Australia, autonomous housing is defined as "housing with completely independent energy, service and waste disposal system" (Parnell and Cole, 1983). A more explicit definition of autonomous house by Vale and Vale (2000) aims to demonstrate that the servicing needs of a building can be met without the use of fossil fuels, and with minimal impact on the environment, by giving control to its users through an alternative technology.

In brief, the idea of resource autonomous is defined as buildings sustainable to reduce the demands of energy and the impacts on natural environment, self-sufficient in the production of their own energy, cleaning the air, grey water and sewage on site, and using healthy materials, for neutral impact, not net positive but also increasing ecological productive and resilience of the natural environment (Salat, 2006; Stasinopoulos, Smith, et al., 2008), as shown in Fig 2–12. Thus, ecological design is generalised by positive interactions with nature, to reduce negative impacts on nature and to increase positive contributions to nature, as shown in Fig 2–13.

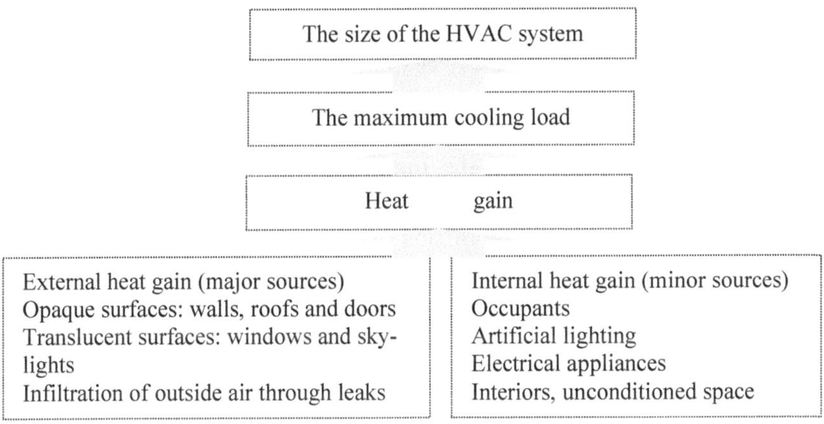

Figure 2–12 Passive techniques to reduce the energy demand in HVAC system

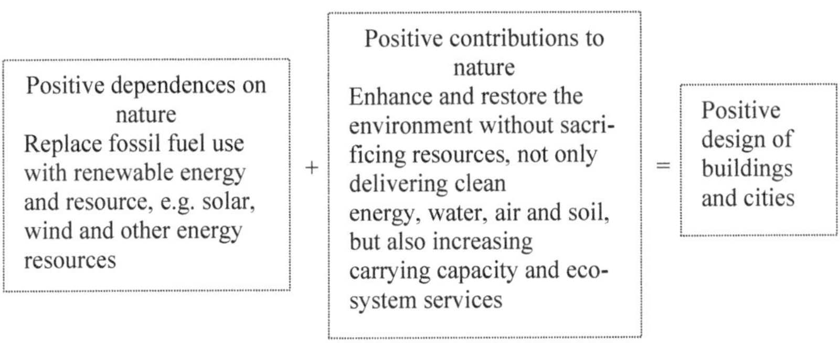

Figure 2–13 Positive design of buildings and cities

2.5.1.4 *Ecological design in a paradigm of nature*

As revealed in modern ecology, biology and thermodynamics sciences, the specialty of nature is an integrated and self-contained living ecosystem; its sustainable prosperity, vitality and resilience are constituted by the perpetual metabolisms of self-organisation, optimisation, and evolution (Mclennan, 2000). In addition, inspired biology science which studies how a biological living organism operates as a self-contained, self-sufficient and self-organising system fit into the niche of nature. They are responsive and adaptive to the natural ecosystem for its ecological sustainable impact, i.e. automatically minimizing their negative environmental impacts upon nature, contributing neutrally and even positively to nature (Portoghesi, 2000). Such paradigms of nature and organisms provide an efficient, visible and constructive pattern for ecological design (Ray-Jones, 2000), in contrast to the linear industrial pattern, in which nature is considered as supporter and sinker for energy and emission fluxes from buildings and cities. In brief, by applying the intelligence of the natural world, by activating the ecological interactions of buildings and cities with the natural ecosystem, the metabolisms of buildings and cities will be improved for sustainable symbiosis with nature.

Thus, in an analogy, buildings will be designed like trees and cities like forests (Aalto, 1986: 221; Hagan, 2001; McDonough and Braungart, 2002), with structure, material and energy associations to meet the demands of architectural and urban utility, function, structure and form (Gauzin-Müler, 2002). For example, it can be realised by using building materials recycle, either directly into another building or for other uses the building's material impacts; to adapt construction and design to the regional climate so that buildings will have the lowest energy use during the operational phase while ensuring ecological sustainability by minimal disturbance on ecosystem and maximal conservation of energy and natural resources. For example, an ecological urban design scheme is generalised in Fig 2–14. The intensive demands of energy and natural resources in cities are induced by various neighbourhoods, buildings and open spaces, their infrastructures and their functional, transport and communication systems. To achieve a closer integration with the overall ecological balance with nature, these interactions are subject to a constant modification and reconstruction following the principles found in natural world (European Charter for Solar Energy in Architecture and Urban Planning).

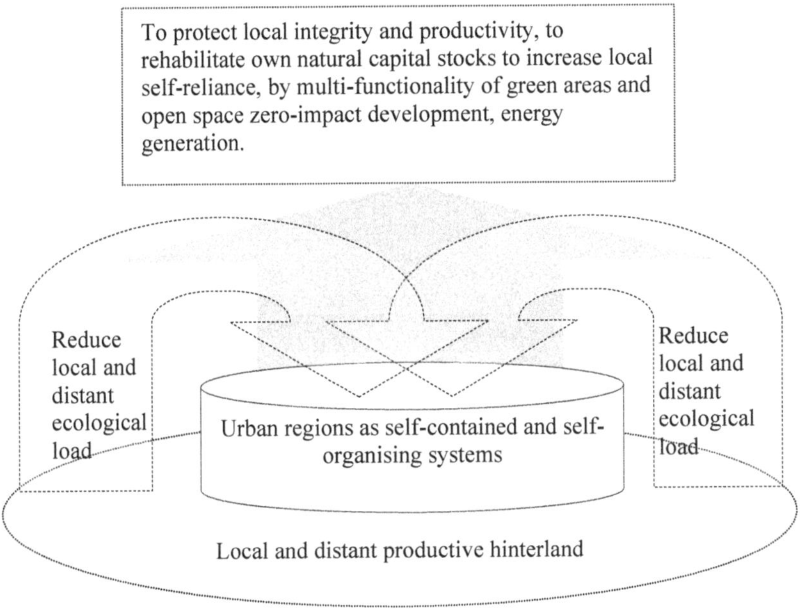

Figure 2–14 A conceptual proposal of ecological urban design

2.5.2 Ecological design in practice for sustainable environment

In principle, most of the current ecological design practices aim at energy saving by applying the principles of ecological design (Ryn and Cowan, 1996), to achieve high-performance, energy efficient and economically viable buildings and cities, and enhance occupants' well-being (LEED). Accordingly, technological and technical innovations in ecological design include innovative construction techniques, ecologically sound building materials and environmentally sensitive energy management and resource conversation, integrated with specific aesthetic contents.

2.5.2.1 High-tech and low-tech in ecological design practice

In general, there are two main strategies for environmentally responsive design in terms of energy and material use. One is advanced engineering and environmental technology, another is historic indigenous methods and materials, which are classified as low-tech and high-tech (Daniels, 2000). In the early of 1970s, following the first oil shock, environmental alternatives were proposed by some architects to use low energy materials, naturally warm, light and easily-worked materials such as timber and earth, an approach based directly on traditional models. The most notable exponent of the low tech remains Paolo Soleri, "arcology" which means architecture consistent with ecology (Steele, 2005). The use of low-tech can risk a tendency towards cliché, without sufficient integration into the contemporary urban environment.

Meanwhile, in contrast to Arcadian minority intent to returning building to a pre-industrial, a rationalist majority interested in developing the techniques and technologies of contemporary environmental design (Hagan, 2001). High-tech architecture is symbolized by the commercial buildings using steel and glass structure additionally powered by solar energy. In addition, the environmentally favorable features and construction technology, like thermal glass, solar collectors, photovoltaic panels, air filtering systems, and recycled construction materials are used to enhance the building-as-art statement (Birkeland, 2002; Morison and Hes, et al., 2006). Since 1990s, the landmark buildings of advanced technology to achieve sustainability, eco-tech, are the Commerzbank tower in Frankfurt and the dome of Reichstag in Berlin, both designed by Fosters. However, the use of high-tech in architecture is not always convincing, particularly in the areas of temperature control in summer and energy-saving in winter.

2.5.2.2 Three cases of contemporary ecological design in buildings

Three design projects of ecological buildings, ranging from commercial, public and to municipal buildings in UK, Singapore and Australia, are briefly reviewed in the following. It aims to draw a general profile of contemporary ecological design practices, including the objectives, strategies, techniques, evaluation, and existing barriers.

The first case is London's first ecological tall building, 30 St Mary Axe[18], Swiss Re Headquarter in London, UK, commissioned by Swiss Re, one of the world's leading reinsurance companies. It is designed by architects Foster and Partners from 1997 to 2004. Stated by the principal designer Norman Foster, it is an embodiment of the core values about humanising the workplace, conserving energy, democratising the way people communicate within a building, and the way that building relates to the urban realm. This building was awarded the Stirling Prize for the best new building in 2004 by the Royal Institute of British Architects, and the 2003 Emporis Skyscraper Award for the best skyscraper in the world completed that year.

The ecological strategies of the building start with the bioclimatic envelope design from the concern of urban context, and then feature with the inner space organisation. The original ecological concept is an analogues model in wind tunnels looking at air flow patterns to enhance the building's environmental performance, as shown in Fig 2–15.

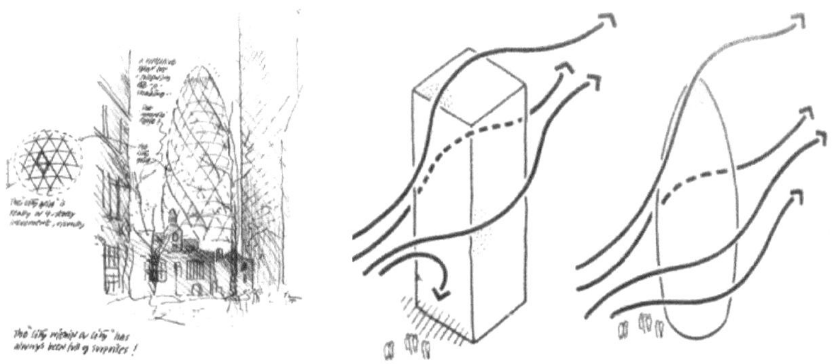

Figure 2–15 Sketches of the wind conditions near the building[19]

As a positive response to the constraints of urban context, the site, the distinctive form of building is slender than a rectangular block of equivalent size. It is generated by a radial plan, with a circular perimeter, the building widens in profile as it rises and tapers towards its apex (ibid). Within the inner space, the tower explores the conceptual

18 2005, Catalogue: Foster and Partners. Munich; London, Prestel

19 Images sources: www.building.co.uk

idea which was theoretically originated by Buckminster Fuller, suggesting a new rapport between nature and the workplace, its energy-conscious enclosure resolving walls and roof into a continuous triangulated skin.

The environmental success include the reduction of reflections and the improvement of transparency; the slimming of its profile towards the base maximises the public realm at ground level, reducing the amount of wind deflected to the ground compared with a rectilinear tower of similar size. It further helps to maintain pedestrian comfort at street level, and creates external pressure differentials that are exploited to drive a unique system of natural ventilation into the building' inner space. In result, the system reduces the towers reliance on air conditioning which is up to half the energy consumed by conventional air-conditioned office towers.

The second case of ecological design is the New National Library Building[20] in Singapore, featuring with low energy design modes and green material selections and specifications. It is designed by TR Hamzah & Yeang from 1998 to 2004. This building presents the library culture as being more fun, an urban motivator for civic activities. It was awarded the platinum Green Mark Award from Singapore's Building and Construction Authority in 2005; and ranked top in the "Energy Efficiency and Conservation Best Practices Competition for Energy Efficient Buildings: New and Existing" awarded by the Association of Southeast Asian Nations (ASEAN) Centre for Energy in 2007.

These are assembled together into a collective strategy for low energy and high comfort, which includes three modes of active, passive and mixed mode of energy use through the building. Specifically, the passive design modes strategies used are: day lighting, good solar orientation and configuration, sun shading, natural ventilation, façade design, building colour and landscaping. The active mode, full air-conditioning and artificial light is operated in most of the library workspaces and collection areas. Into this blend of the active and passive, a third strategy mixed mode is adopted, where natural ventilation

20 It is 15 storeys, 98m high public building, commissioned by the state of Singapore. http://www.llewelyn-davies-ltd.com.

is supplemented by mechanical means such as fans, in central transition spaces, for example, lobbies, foyers, courtyard. These are supported by airflow and energy simulations, along with predictions of comfort.

In this ecological building, ecological design approach through the building also involves the holistic consideration of the careful selection of materials used in the designed system and the endeavour by design to reduce the impacts of this use on and its integration with the natural environment, over the life-cycle of the designed system. The intended use of recycled and reused materials generally such as the carpet wall fabrics and sustainable-forested local timber will reduce impact considerably.

As a positive outcome of significant attention paid to the key factors in lowering energy embodiment and lowering environmental impact in the building as compared to the typical office building type, the environmental impact of the building is lower as compared to a typical office building. Assessed in terms of "Energy Embodiment Analysis of the Building", the total energy embodiment was calculated to be 17 GJ/m2 which is within accepted benchmarks for commercial buildings, i.e. between 10 – 18 GJ/m2. The building performance of this ecological building is illustrated by virtue of the simulation conducted on the energy consumption, around 185 KWH/m2/annum, far more efficient than a typical commercial office tower of 210 KWH/m2/annum in Singapore.

The third ecological building is Council House 2 in Melbourne, Australia[21]. It is designed by Advanced Environmental Concepts, Lincolne Scott, DesignInc. It provides healthy and productive workplace for the occupants, staffs in City of Melbourne, while reducing the building's impact on the environment through comprehensive approaches of ecologically sustainable design. It is rated as Australia's greenest and healthiest purpose built office building, Six Green Stars from Green Building Council of Australia in 2006, assessed by the Green Star evaluation method of the environmental performances, i.e.

---

21 Council House 2 is a 10-storey office building for about 540 City of Melbourne staff, with ground-floor retail spaces and underground parking, officially opened in August 2006. http://www.melbourne.vic.gov.au

the use of water, materials, transport innovation, emissions, land use and ecology energy, management and indoor environmental quality, as shown in Fig 2–16. It is also a United Nations award-winning building.

It has comprehensive sustainable technologies incorporated into every conceivable part of its 10 storeys. Despite the normal ecological design principles, i.e. using thermal mass for cooling, using plants to filter the light, there are some sustainable innovation in this ecological design approaches. A water-mining plant in the basement, phase-change materials for cooling, automatic night-purge windows, wavy concrete ceilings, a façade of louvres powered by photovoltaic cells that track the sun, even the pot plant holders have involved a whole new way of thinking. Compared with the existing Council House, Council House 2 reduces its electricity consumption by 85 per cent and its gas consumption by 93 per cent, which means CH2 will use only 13 per cent of the energy consumed by the existing Council House; thus CH2 emissions will be 20% of the emissions of the current Council House.

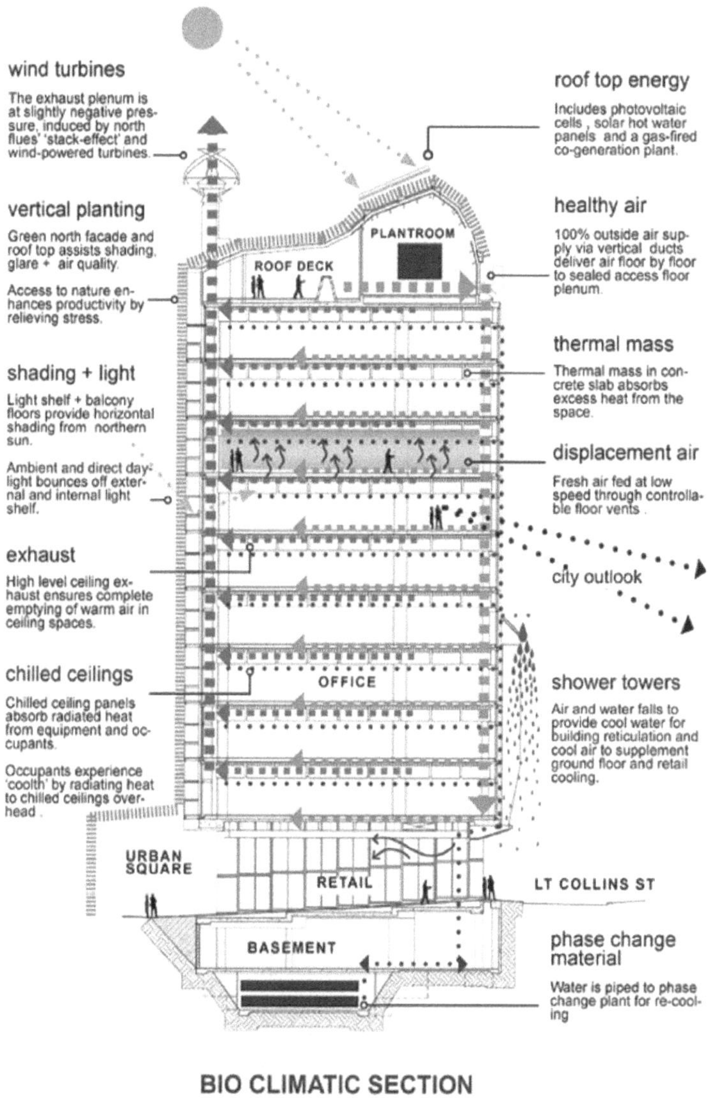

Figure 2–16 Comprehensive ecological technologies in Council House 2 Melbourne[22]

---

22 Image source: C40 cities climate leadership group www.c40cities.org. Copyright permission of diagram is granted for the reproduction in this book by C40CITES, Climate Leadership Group, UK, and the City of Melbourne, Australia.

A comprehensive eco-audit of the materials used in CH2 will assess all aspects of the manufacture and transportation of materials in relation to their effect on the environment and the occupants of the building. Although the reduction in energy costs will be substantial, the greatest economic benefit is expected to be in increased productivity, reduced absenteeism and lower staff turnover rates, which cost employers millions of dollars each year. Studies have shown the improved air quality of systems like CH2's air conditioning system could achieve a 4.9 per cent increase in productivity, in part through reduced sick leave.

In brief, in the case of Foster's strategies for ecological building are active responses, externally from a perspective of wind pattern study influencing building's form and internally from a concept of combining green spaces into building inner spaces as naturalism idea; while Yeang's are effective facilitations by mix modes of natural and artificial ventilations as inspired typically in subtropical regions, and additionally specified with selections of green materials; in the case of CH2, ecological design strategies are a comprehensive combination of all the current available specific ecological techniques and technologies into one project.

2.5.2.3 Evaluation of ecological design practices

The environmental performances of ecological buildings and cities are majorly measured by the degree of investment in energy-saving systems, the durability of construction materials, and the number of recycled products used in fabrication (Salat, 2006). For example, life cycle analysis (LCA) is a classical approach assessing the economics of buildings, including the initial capital cost of construction, the operating costs of utilities and maintenance. Some environmental measurement criteria have been developed worldwide, for example, British Research Establishment Environmental Assessment Method system (British Breeam), the North American Leadership in Energy and Environmental Design system (LEED), and Green Star from the Green Building Council of Australia. They are constituted to evaluate the environmental initiatives of designs, projects and buildings in terms of energy and water efficiency, indoor environmental quality and resource conservation.

2.5.3 Application of computer science in sustainable design

However, in the current ecological design practice, there are still several technical and technological barriers challenging the realisation of the conceptual proposal of ecological design. For example, active ecological interactions between buildings and the natural ecosystem in the forms of material resources and energy uses over their life-cycles, to realise zero or neutral ecological impact and even positive contributions, including biodiversity loss and habitat degradation, have not been fully manipulated quantitatively into practice yet. One of the main problems in architecture today is recognized as the quantity of information and the level of complexity involved in building projects, especially in high-rise and large-scale urban projects (Vrachliotis, 2008). In this background, the emergence of complexity science, cybernetics has conjured up as a structural-scientific foundation of which the technological potential and cultural magnitude for architectural production can be discussed only from the perspective of contemporary information technology. In brief, in the Age of Information, computer-aided-design has been introduced into building industry, accelerating the development of a new visionary eco-digital architectural iconography and expanding theoretical instrument in architectural-theoretical discourses (Steele, 2005; Vrachliotis, 2008).

2.5.3.1 A brief history of digital architecture and urban design

Since the early of 1970s, there has been a wide interest in employing advanced digital modelling techniques to better address the lack of suitable design methods and tools to constitute higher complexity of urban environment. These digital data modelling systems follow from a history of urban modelling and simulation, which includes the work of Land Use and Built Form Studies (LUBFS) at Cambridge University (Baxter, Owers and Echenique, 1975), and geographic information systems (GIS) based systems, i.e. the London Digital Model developed by Batty (2006). More recent suggestions include the use of advanced building information models (BIM) at Gehry Technologies (Shelden, 2009), and their link to energy and performance analysis programs to provide more powerful tools as they are aggregated. Some research groups, i.e. Fournier at University College London in UK, Frazer's Groningen experiment in Netherlands (Frazer, 1996), and Schumaker (2006) at Zaha Hadid's Architects in UK, are making proposals for the generative use of these techniques to sustainable design.

2.5.3.2 Information modelling of buildings and cities

Digital modelling guides the operational procedures and policies in architecture and urban design, with quantitative and qualitative criteria for the acquisition and processing of information and methodological and operative procedures (Carlo, Albisinni, et al., 2006). Over the years, it is possible to realise necessary updating and monitoring. The nature of urban design, conventionally characterised with complexity and chaos, is possibly facilitated in compactness for prospective sustainable design and development. For a basic comprehension of current information models in design practice, approaches to digital modelling of buildings and cities are generalised as follows:

- A framework of the environmental program, which integrates geo-spatial information, building information and other morphology of built environment (Jemtrud, Nguyen, et al., 2006);
- Matrixes of geometric and topographic parameters, which is logic flows of encoded issues of urban and architectural conditions (Carlo, Albisinni, et al., 2006);
- Virtual environment of thematic data of urban environment, i.e. noise, flooding data by modelling and rendering techniques, spatial algorithm of culling, simplification, and normal mapping and image caching (Carlo, Albisinni, et al., 2006);
- A semantic digital system with simulation techniques supplied by administrative data, i.e. information about planning regulations, planning process management, public participation (Friedrich, 2006).

With the establishment of these fundamental spatial dataset, a virtual information model of urban spatial system corresponds to the physical urban reality with hierarchic order. It allows a consistent representation of the built environment, and identifies the compositional and constructive aspects of urban fabric. In computer techniques of iterative performance evaluation, evolution and optimization, environmental management of buildings and cities can be fully analysed. Specifically, the dynamic and complex interactions between different variables start from the initial data collection and integrated process; then, computer information process is used to evaluate the efficiency of the input consumption, to optimise the environmental performance of buildings and cities, and to finally minimizing the emission from the consumption (Caldas, Norford and Rocha, 2006), as shown in Fig. 2–17.

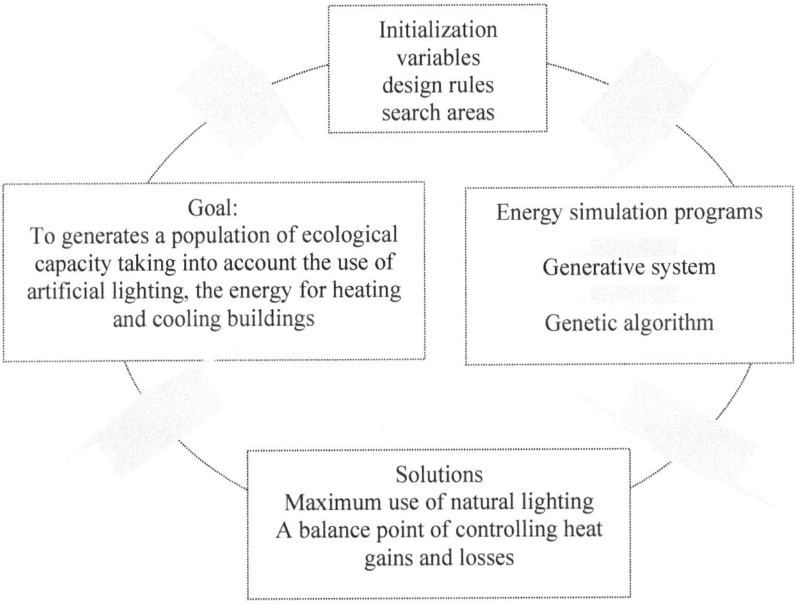

Figure 2–17 A specific generative digital building modeling on energy concern

2.5.3.3 *Impacts of digital design in sustainable environment*

The emergence of digital design implies a novel idea to consider sustainable design as a dynamic interactive process to achieve a desirable consensus in a generative and evolutionary approach (Frazer, 1995). More significantly, the computer interface facilitates active engagements and participations of not only the designers, but also the end-users of buildings and cities into the information communication channel (Hagan, 2001), for example, the instant evaluation of the impact on the environment and human well-being. Therefore, positive interactions of the environment and the end-users can be accessible in a feedback control loop for adaptation and optimisation towards sustainable environment, which brings self-awareness into sustainable design, as shown in Fig 2–18.

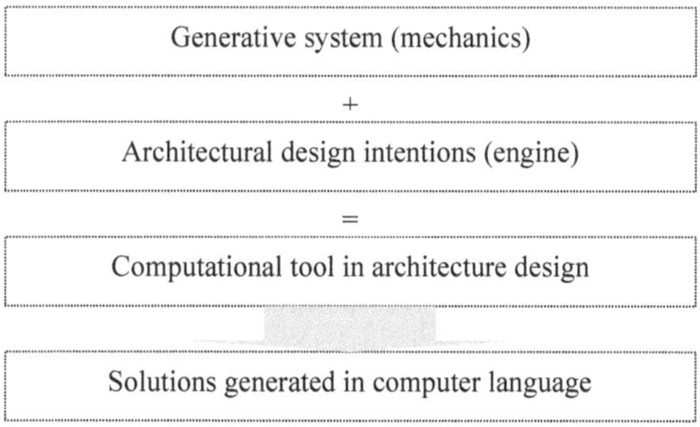

Figure 2–18 Generative digital modeling in architecture design

For example, to analyse the full participation of variables of thermodynamic performance of HVAC system in life-cycle of the buildings, computer simulation models have been developed as analytic tools for buildings with some sustainable certifications. LCADesign is developed by CSIRO in Australia for ecological analysis of building environmental impact, Ecotect by CRBE in UK for analysing a building's performance with an emphasis on energy efficient and sustainable designs, TAS by EDSL in UK for the thermal simulation of buildings, EnergyPlus by DOE in USA for energy simulation, BEES by EPP in USA for measuring the environmental performance of building products, and THERMOPTIM by CSTB in France for calculating the complex thermodynamics of buildings.

In brief, by the technical implementation of computer technology in digital design for sustainable environment, the key parameters of a design concept, including topography and environmental factors, are translated into a code script (Steele, 2005). It radically changes the conventional design approaches of either bottom-up or top-down, to achieve the vision of sustainable development with integrated respects to social, economical and environmental dimensions (Hagan, 2001). However, in the current digital modellings, an alternative mechanism, optimising environmental performances of buildings and cities in the context of the natural ecosystem and being actively involved by the end-users of buildings and cities, is still missing yet.

2.5.4 A hypothesis: an intelligent model for holistic sustainable design in buildings and cities

Comprehensively, a holistic scheme of sustainable design is constituted by the parameters of the micro-climate demands from the end-users of buildings and cities, the environmental context of natural ecosystems, such as the local climate conditions, available energy and resources and prevalent building technologies. To realise the sustainable symbiosis of buildings and cities with the natural ecosystem, e.g. reduced negative environmental impacts and even positive contribution into the natural ecosystem, buildings and cities are proposed in the paradigm of nature, to self-organise, adapt and optimise the multiple interactions of all parameters. These interactions are in a non-linear pattern, not the same as the linear industrial pattern without any concern or feedback mechanism for the optimisation of environmental performances. In brief, systematic parametric scheme of sustainable design in buildings and cities are integrated by human dimension, the natural environment, and the availability of building technologies, as three essences, e.g. man, nature and buildings or cities.

However, based on the critical discourse of ecological design in theory and in practice in the Age of Information and Ecology, current ecological design strategies and techniques for sustainable environment are fragmented (Stasinopoulos, Smith, et al., 2008), without a complete integration of ecological interactions of buildings and cities with nature. Furthermore, the evaluations of those design propositions are usually post-performance based, without predictable and adaptive control mechanism for the necessary on-going optimisation over the life-cycle of buildings and cities.

Technically, these complex ecological interactions between buildings and cities with the context of natural ecosystems, with the involvement of human environmental concern, can be considered as on-going environmental performances of buildings and cities. In information modelling, all these interactions can be represented, simulated, monitored, computed, predicted, controlled and optimised in feedback loop by the information communication technology. In other words, with the digital information support, the holistic scheme of sustainable design can be developed as an intelligent model to optimize the environmental performance of buildings and cities, engaged by the end-users and integrated within the context of natural ecosystems, for minimal, neutral and even positive ecological impacts upon the natural ecosystem.

Thus, an intelligent model is proposed for holistic sustainable design in buildings and cities, which is configured and parameterized with a holistic scheme consisting of man, nature, and buildings or cities; it is comprehensively supported by information science and technology, i.e. algorithm and techniques to constitute a virtual information system, simulating, computing and optimising the environmental performances, for sensitive and adaptive interactions with natural environment, while meeting the environmental demands of human activities, and minimise negative impacts and maximise positive impacts, as shown in Fig 2–19.

This intelligent model of sustainable design also reflects the new world-view of the Post-Modernism, i.e. the creativity of the universe is evolved by self-organising in feedback to generate order or organised complexity. To explore such an intelligent model with the aesthetics of universal generality, a scientific framework revealing the fundamental relationships of man-made environment with nature, both macroscopic and microscopic mechanisms is essential (Portoghesi, 2000; Masanori, 2000).

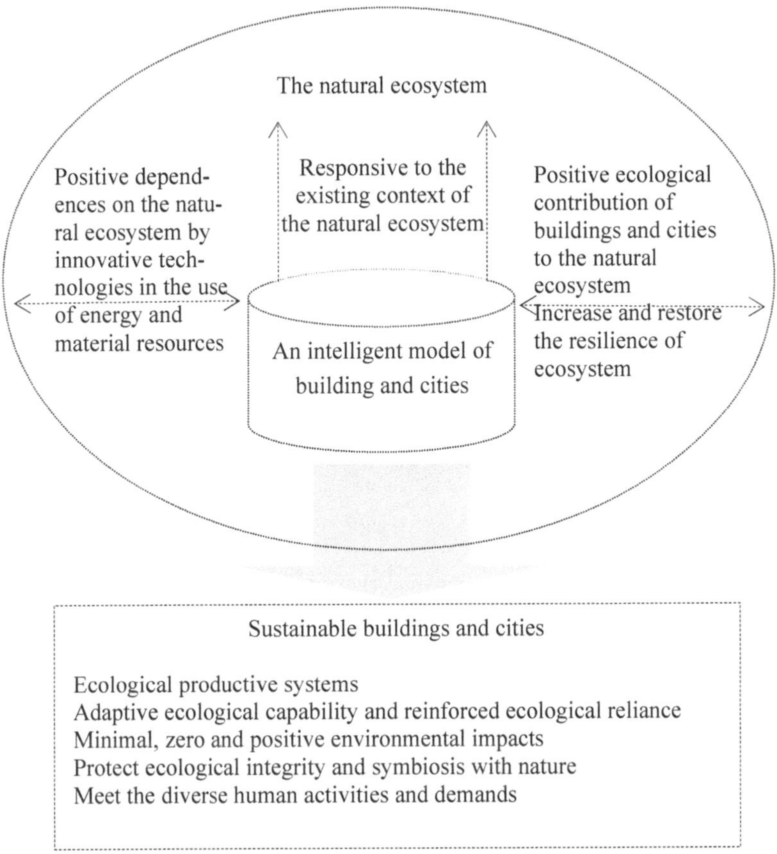

Figure 2–19 A hypotheses: an intelligent model of holistic sustainable design

In summary, contemporary theories and practices of sustainable design in the built environment, including the strategies, propositions and technologies, in the Age of Information and Ecology since the mid-20th century, have been empowered by modern progresses in ecology and biological sciences, which are both ecologically responsible and symbolically communicative (Prigogine and Stengers, 1984; Wines, 2000; Jencks, 1995; Gleiniger and Vrachliotis, 2008). In this background, a hypothesis of sustainable design is a holistic sustainable design scheme embracing the dimensions of man, nature, buildings and cities: adapting and optimising environmental performances of buildings and cities to effectively reduce their negative impact or even contribute positive impact on the natural environment, which is also human accessible via information communication technology. In brief, an alternative paradigm of sustainable design will integrate both ecology design philosophy and information science.

Part II
The Framework of Open Systems Evolution

Chapter 3
The model of open systems evolution

To meet these challenges of sustainable development, it is essential to explore new theories and methods, frameworks and paradigms for design in the built environment towards sustainable environmental symbiosis of buildings and cities with nature. Especially, a scientific framework to establish sustainable ecological relationships of buildings and cities within the context of the natural environment is considered as essential.

In natural science, thermodynamics laws are one of the fundamental laws generalising the mechanisms of energy use and transfer in thermal systems. Classical formulations of the Second Law of Thermodynamics introduce the irreversibility of entropy in isolated systems, such as degradation of energy, waste and unavailable energy during thermodynamic transformations. However, a striking modern statement of the Second Law to open systems states not only the irreversible entropy but also the emergence of order as essential consequences of the open system's evolutionary thermodynamics. In brief, an open system evolves as a dissipative structure from a far-from-equilibrium state to a non-equilibrium stable and compatible interrelation with the host environment, by re-distributing material and energy within the system. This open system model provides as a scientific framework to interpret, analyse and optimise environmental performances of buildings and cities within the context of the natural environment, in the terms of thermodynamic exchanges, including both energy and resource consumptions and environmental impacts upon the natural ecosystem.

3.1 The Second Law of Thermodynamics

Originally, the term of thermodynamics, coined from Greek words for heat and movement, was originally concerned with steam engines for the efficient conversion of heat to mechanical work demanded in the Industrial Revolution (Prigogine and Stengers, 1984; Von Baeyer, 1998; Schneider and Sagan, 2005; Carnot, 2005; Sandler, 2006; Huber and Mills, 2006; Stowe, 2007). Thus, the science of thermodynamics is developed as a combination of physics, chemistry, biology and engineering (Von Baeyer, 1998), providing qualitative and quantitative formulations of energy transformations in properties, such as pressure, volume, temperature and composition (Roy, 1995; Sandler, 2006). In biological and chemistry thermodynamics, for example, it is recognised that energy can be converted in chemical reactions from matter, and matter can be converted to energy in bio-chemical reactions (Prigogine and Stengers, 1984; Casti, 2000; Moore, 2006). The fundamental behaviours of thermal particulars has been generalised in thermodynamics laws (Stonier, 1990; Roy, 1995): the First Law on the conservation of energy in isolated systems, the Second Law on spontaneous change of thermodynamic transfers of open systems, the Third Law on the unattainability of absolute order of crystallisation of matter at absolute zero temperature.

In the thermodynamic laws, a thermodynamic system is formed by the contents of defined geometric volumes of macroscopic dimensions in a context of surroundings, which is termed as a host environment (Prigogine, 1961). A complete configuration of thermodynamic transformation consists of a thermodynamic system and a host environment, as shown in Fig 3–1.

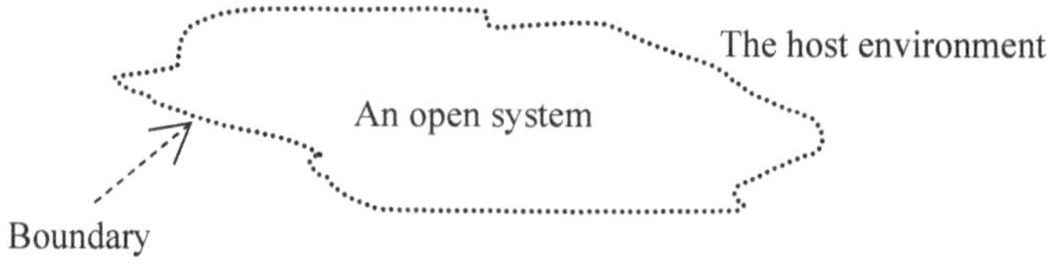

Figure 3–1 Configuration of an open system in the host environment

In terms of thermodynamic exchanges across the boundary, systems are classified three types as follows (Prigogine, 1961; Prigogine and Stengers, 1984; Adami, 2002; Schneider and Sagan, 2005):

- An isolated system, which cannot exchange energy or matter with the host environment;
- A closed system, which can exchange energy but not matter with the host environment, e.g. the planet in the context of the universe; and
- An open system, which can exchange energy, matter and information with the host environment, e.g. living organic systems in the natural ecosystems (Aon and Cortassa, 1997; Ho and Saunders, 1979).

In each case of thermodynamic systems, thermodynamic interrelations with the host environment vary in terms of the gradients, boundaries, etc. For example, the external gradients between an isolated system and its host environment do not affect the thermodynamic transformation in the isolated system; while the internal gradients between the constitutional components determine the thermodynamic transfer within the isolated system. The thermodynamic transfers of open systems are more complicated than isolated and closed systems, due to continuous fluxes of energy, matter and information across the boundaries of the open system to support the continuous thermodynamic evolution of the open system within the host environment.

Basically, there are two formulations of the Second Law of Thermodynamics. One is the classical statement, generalizing thermodynamic exchanges of isolated and closed systems and emphasizing maximum entropy production as a result of heat death of the systems (Prigogine, 1961; Prigogine, 1978). The second formulation is applicable to open systems, generalizing the mechanism of evolutionary thermodynamics, via which, an open system evolves from disorder to a highly organized complexity, i.e. minimal entropy production along with a compatible thermodynamic interrelation of the open system with the host environment.

3.1.1 Classical formulation of the Second Law to isolated and closed systems

The classical formulation of the Second Law of Thermodynamics was initially studied by an engineer Sadi Carnot on a technology innovation of engine efficiency in 1824; while a formal statement of the Second Law was formulated by German physicist Rudolph Claussius in 1856, based on his physical observations of thermal transfer in isolated and closed systems (Prigogine and Stengers, 1984; Von Baeyer, 1998; Schneider and Sagan, 2005). In general, the Second Law of Thermodynamic connects mechanical and thermal effects into the science of thermodynamics as a non-classical science, quantitatively stating the existence of a universal tendency toward the dissipation of useful energy and the degradation of mechanical energy. Initially, in the 19th century, it has been stated based on the heat transfer of isolated systems: any isolated system inevitably decays, becoming more disorganised, suffering dissipation and finally dying towards the dissolution of entropy; in other words, the total entropy in an isolated system tends to increase over time, approaching a maximum value at the state of thermal equilibrium (Clausius, 1850, 1865, 1968; Kestin, 1976; Roy, 1995; Von Baeyer, 1998; Huber and Mills, 2006). In this transfer of isolated systems, fundamental thermodynamic phenomena consist of gradients, entropy, thermodynamic equilibrium, and etc, which are specified as follows:

3.1.1.1 Degradation of gradients for spontaneous thermodynamics

The first principle of the classical statement of the Second Laws of Thermodynamic is that, heat flows only downhill, from hotter to colder, never uphill (Von Baeyer, 1998; Schneider and Sagan, 2005). It is the gradients of temperature, pressure, density, concentration, and other physical and chemical differences within an isolated system, between an open system and another, or between an open system and the host environment, which forces spontaneously thermodynamic transfer in the direction from higher to lower to extract useful work; the bigger the gap of gradients, the more useful work for extracting (Prigogine, 1961; 1978; Roy, 1995; Moore and Stanitski, et al., 2001;

Schneider and Sagan, 2005; Huber and Mills, 2006; Sandler, 2006; Jungermann, 2006; Atkins and Paula, 2006), as shown in Table 3–1.

Table 3–1 Gradients for spontaneous thermodynamic transfer: downhill (Stowe, 2007)

As two interacting systems approach equilibrium	After they reach equilibrium	The consequence
Heat flows toward the system with the lower temperature	Their temperature are equal	If heat is added, the temperature must rise, $\Delta T\ \Delta Q > 0$
Boundaries move toward the system with the lower pressure	Their pressure are equal	If volume is increased, the pressure must fall $\Delta p\ \Delta V < 0$
Particles flow toward the system with lower chemical potential	Their chemical potentials are equal	If particles are added, the chemical potential must rise $\Delta \mu\ \Delta N > 0$

In brief, the tendency to reduce gradients in temperature, pressure, and chemistry sets up a pre-condition for thermodynamic fluxes and dissipation of energy (Schneider and Sagan, 2005), which is applicable not only to isolated and closed systems, but also to open systems, such as the organisations in economics (Caldarelli, 2004), chemical reactions, ecosystems and solar systems. Biological systems, for example, are the most successful at searching for gradients, capturing these gradients, degrading them, and improving their perception, intelligence and capacity, evolving into organised complexity (Schneider and Sagan, 2005). The greatest gradient in the universe is the solar gradient; it is the energetic difference between sun and the earth which provides the condition of global thermodynamic equilibrium-seeking systems and helps maintain the climate in which life thrives.

3.1.1.2 Dissipation of energy for irreversible entropy

Observed in the thermodynamic transfers of isolated systems, it is generalised one could not finish it with the same amount of energy as which one started, once energy is expended, changing it from a usable to an unusable form, it can't be replaced, it is dissipated. In other words, the state of an isolated system inevitably decays from more ordered to less. In the modern time, the thermodynamic transformation of isolated systems is interpreted more precisely from physical-chemistry perspective, which states, in an isolated system, the internal portions at different temperatures will spontaneously disperse and dissipate, and original energy will be degraded into low quality of forms (Prigogine and Stengers, 1984; Nicolis and Prigogine, 1989; Lambert, 2002; Atkins, and Paula, 2006). In the case of closed systems, during the thermodynamic transfer from one system in higher grade to the one in lower, high-grade energy starts as the inputs is consumed in the process of structuring, organizing and moving to increase the order of the system, and then some fraction is discarded in the form of low-grade heat as the outputs of transfer.

3.1.1.3 Maximum entropy production as "the arrow of time"

During the thermodyanmcis of isolated and closed system, some energy is irreversiblly lost and waste, unavailable for the convertion to effective mechnical work, which is called entropy, normally termed as disorder of isolated system (Schrödinger, 1962; Schneider and Sagan, 2005; Sandler, 2006). As a state function, entropy's variation, in the transition of a system from a state to another, can be calculated. Its production can be written as a sum of the products of generalised forces or affinities and the corresponding rates or fluxes of the irreversible processes, increasing maxmumally at the state of thermodynamic equlibirum (Prigogine, 1961; Prigogine, 1978; Prigogine and Stengers, 1984). The irrevsibility of entropy in thermodyanmic transformation in isolated and closed systems is normally called as "the arrow of time", an evidence of the existence of time, giving a direction to the succession of states of a system (Eddington, 1958; Prigogine and Stengers, 1984; Nicolis and Prigogine, 1989; Prigogine, 1991; Schneider and Sagan, 2005).

3.1.1.4 Thermodynamic equilibrium in isolated and closed systems

The thermodynamic transformation in isolated or closed system, due to gradients, will result in the loss of the differentials and adjusting to a single uniform temperature, concentration and etc. It means all the initial gradients, e.g. the quantities of matter, the pressure differences, density differences, and temperature differences between the thermodynamic system and the host environment in the case of closed system, or between the constitutional components within the isolated system, tend to equalize over time, which is termed as thermodynamic equilibrium (Schrödinger, 1962; Nicolis and Prigogine, 1989; Roy, 1995; Sandler, 2006). In brief, at the thermodynamic equilibrium state, no more gradients for further thermodynamic transformation in isolated systems or closed systems, after some useful energy being converted to effective work, i.e. order is upgraded in the systems (Kay, 1984).

3.1.1.5 Summary

In summary, there are three consequences of thermodynamic transfers in isolated and closed systems, as generalized in Tab 3–2. The first is thermodynamic equilibrium, the uniformed gradients to finish thermodynamic transformation; the second is order being upgraded in the lower-grade system by useful work; the third is entropy irreversibly produced by the systems. In brief, the reduction of gradients and the dissipation of energy towards thermodynamic equilibrium result in the growth of order and the production of entropy in parallel.

Table 3–2 Comparison of thermodynamic transformation in isolated & closed systems

Thermodynamic transformation	Isolated systems	Closed systems
Gradients force for spontaneous change	Internal gradients between the constitutional components	External gradients between one system and another
Thermodynamic process	Dissipation of physical and chemical differences, e.g. heat, temperature, concentration and etc, as useful inputs	
Consequence	Thermodynamic equilibrium: uniform of temperature and concentration within the isolated system; no more thermodynamic transformation will take	Thermodynamic equilibrium: uniform of temperature and concentration between the systems; no more thermodynamic transformation will take place between the closed system and another or its host environment

	place within the isolated system	
	Order: organized complexity is upgraded in the system	
	Entropy: some energy is unavailable for useful work	

3.1.2 Modern formulation of the Second Law to open systems

3.1.2.1 Contradiction of heat death and universal growth of organic lives

In the real world, the biological system on the earth exchanges energy with the universe, i.e. solar energy constantly from the sun, which is generally considered as a closed system (Prigogine, 1961; Schneider and Sagan, 2005; Smil, 1999; Huber and Mills, 2006), as shown in Fig. 3–2. In terms of the classical Second Law of Thermodynamics on closed systems, the irreversible consequence of thermodynamic exchanges between the earth and the universe is thermodynamic equilibrium in maximum entropy by degrading solar energy (Bertalanffy, 1968; Sandler, 2006; Hershey, 2009). Thus, a hypothesis is the irreversible tendency towards a disorder or "heat death of the universe", a state in the absence of structure, the loss of information and usable energy (Roy, 1995; Prigogine and Stengers, 1984; Nicolis and Prigogine, 1989; Stonier, 1990; Persall and Trumble, 2003), which would result in a destruction of diversity in biological systems, ecosystems and especially human systems.

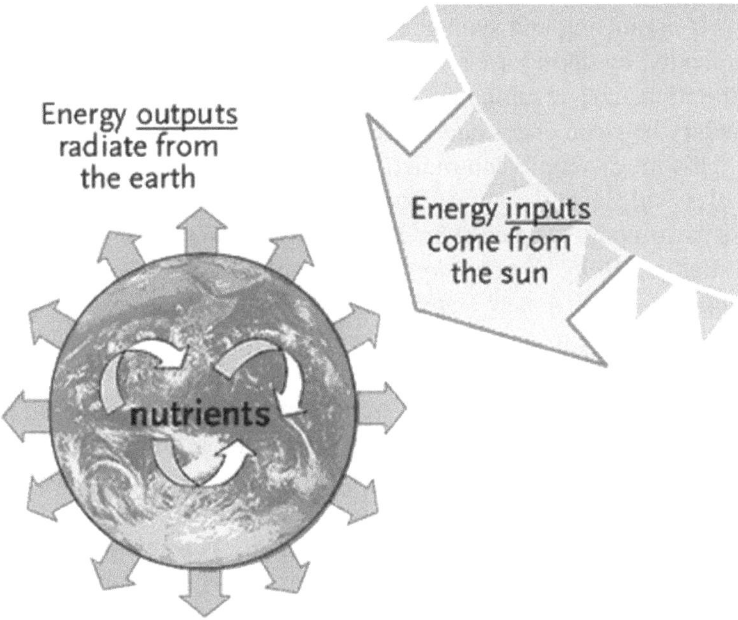

Figure 3–2 Thermodynamic exchange between the earth and the sun[23]

However, what is observed on earth is a successive diversification of life and entropy continually decreasing (Prigogine, 1980; Prigogine and Stengers, 1984; Prigogine, 1991; Wright, 2000; Davies, 2004), which seem contradictive with the heat death hypothesised by the classical Second Law. This is true not only of biological systems, but it becomes even more apparent in cultural evolution, technological evolution, and the evolution of human information systems (Bertalanffy, 1968; Delanda, 1997; Davies, 2004; Huber and Mills, 2006). Living organic systems in the natural ecosystem are thus argued not as closed or isolated systems, but open systems, to which the classical Second Law of Thermodynamics is not applicable to (Prigogine and Stengers, 1984; Nicolis and Prigogine, 1989). In fact, living systems

23 Diagram source: Forest Ecology and Worms: How Forest Ecosystems Work, www.nrri.umn.edu. Copyright permission for the reproduction in this book is granted to the author by Great Lakes Worm Watch Program, Natural Resources Research Institute, University of Minnesota Duluthat.

are self-organising and spontaneous growing towards higher levels of complexity, by taking use of negative entropy, e.g. energy, matter and information, and resulting in organised complexity emerging at the boundary between chaos and order (Schrödinger, 1962; Stonier, 1990).

This mega-scopic contradiction leads to expand the limitation of the classical Second Law of Thermodynamics on isolated and open systems to open systems (von Bertalannffy, 1968; Hemmer, Holden and Ratkje, 1996; Katchalsky and Spangler, 1968; Glansdorff and Prigogine, 1971). A new statement of the Second Law of Thermodynamic applicable to open systems is termed as irreversible thermodynamics or non-equilibrium thermodynamics, and evolutionary thermodynamics, which introduces the significance of non-equilibrium as a source of order and dissipative structure for the reformation of material and energy distribution (Prigogine, 1978; Prigogine and Stengers, 1984). It contributes a microscopic formulation of an evolutionary pattern in microscopic level, encompassing all of chemistry and essential parts of biological and ecological thermodynamics, as well as for many other complex systems sciences such as meteorology and geology (Prigogine, 1961; Brooks and Wiley, 1988).

Hence, the contradiction of the universal growth of organic lives on the earth and the heat death of the universe can be interpreted in such a way: the gradients of solar energy from the universe initiate the thermodynamic exchange of organic lives as the open systems on the earth, which are observed as the growth and evolution of the natural world; meanwhile, irreversibly entropy increases in the natural world. In brief, there are two arrows of time in the evolution of the universe as an open system: one is entropy production into the universe; another is the growth of the organic lives on the earth.

3.1.2.2 Open systems as dissipative structure

The evolutionary or irreversible thermodynamic exchange of open systems with the host environment is more complex than the thermodynamic transformations of isolated and closed systems, as shown in Fig 3–3. An open system spontaneously exchanges with its host environment, driven by the physical and chemical gradients such as temperature, pressure, density differentials between them; continuous input fluxes from the host environment consist of energy, matter and information called as neg-entropy (Schrödinger, 1962) sustain the evolution of the open system.

In evolutionary thermodynamics, an open system is termed as a dissipative structure, the order of which is achieved at the expense of disorder in the host environment, which is usually called order out of chaos. By degrading gradients and dissipating matter and energy from the host environment, an open system evolves from disorder of a nonlinear far-from-equilibrium state to order of a non-equilibrium state, when organized complexity emerges with a highly organized structure of energy and material resources within the open system (Prigogine, 1961; Nicolis and Prigogine, 1977; Prigogine and Stengers, 1984; Nicolis and Prigogine, 1989; Schneider and Sagan, 2005; Marchettini, Pulselli and Tiezzi, 2006), as shown in Fig. 3–4.

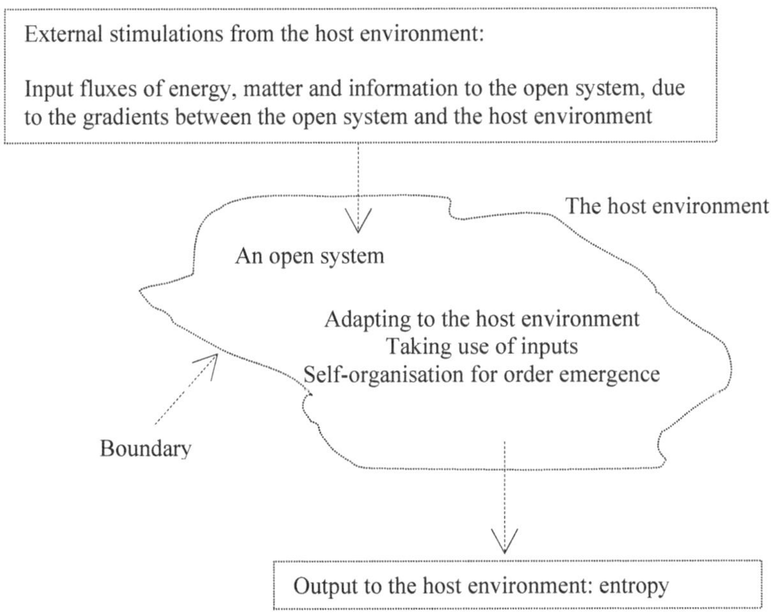

Figure 3–3 Evolutionary thermodynamics of an open system in its host environment

The evolutionary formation of matter under the constraints imposed from the host environment distinguishes dissipative structure to conservative structure in classical science (Nicolis and Prigogine, 1989). Example of dissipative structures in nature can be found in hydrodynamic systems and living organism systems (Prigogine, 1978; Schneider and Sagan, 2005). Living biologic organisms dissipate energy and matter to support their steady-state operations and the production of entropy, i.e. increasing disorder in the surrounding and increasing the organization of the living organic systems.

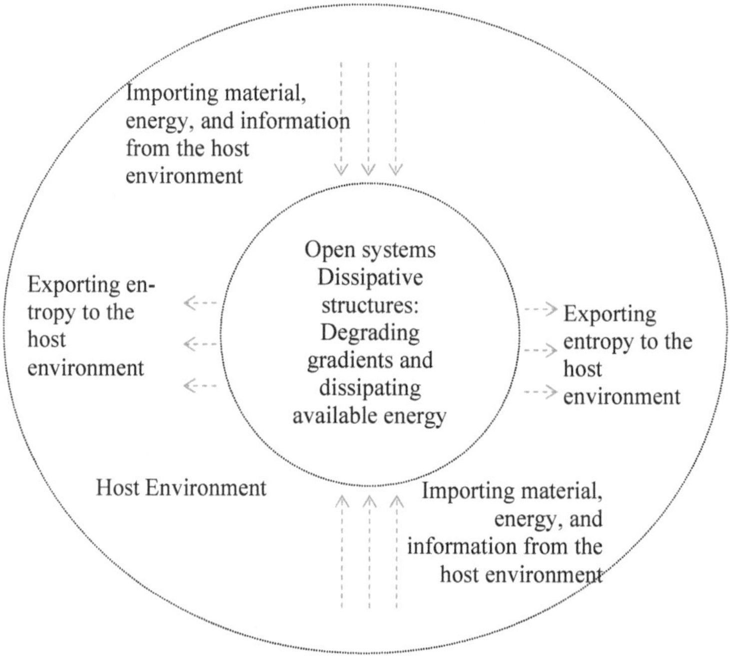

Figure 3–4 Evolutionary thermodynamics of an open system as dissipative structures

3.1.2.3 Transitional phases for self-organization

From the internal microscopic level, with the fluxes of energy and matter across the boundary, the state of an open system evolves through several transitional phases, including equilibrium, near-equilibrium, non-linear far-from-equilibrium, and non-equilibrium (Prigogine, 1961; Prigogine and Stengers, 1984; Schneider and Sagan, 2005; Sandler, 2006), as shown in Fig 3–5.

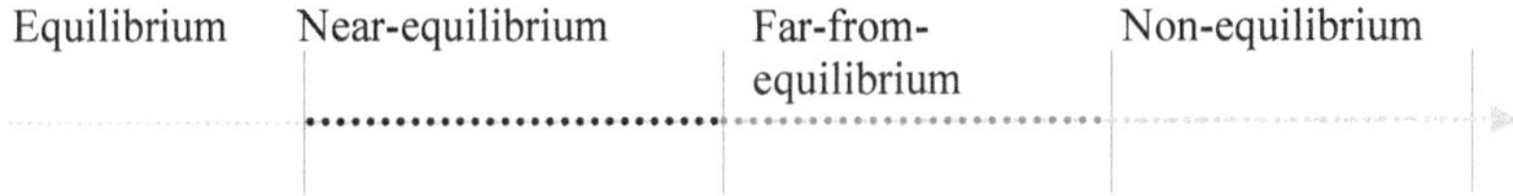

Figure 3–5 Transitional phases of open systems evolution

- Phase I of equilibrium: entropy is zero, no further changes can be expected within the system since no application of gradients;
- Phase II of near-equilibrium: continuous gradients as thermodynamics applies as driving forces while they are weak, the rates are linear functions of the forces, predictable behaviour is observed;
- Phase III of far-from-equilibrium: predicting behaviour is more difficult, the nonlinear response and relationship prevail, any small perturbations or fluctuations can become amplified into gigantic, structure-breaking waves (Prigogine and Stengers, 1984);
- Phase IV of a stead-ready state of non-equilibrium or a stationary non-equilibrium state: a highly ordered organisation in the open system and a compatible interrelation of open system with the host environment are achieved (Prigogine, 1961; Kay, 1984, 2002; Prigogine, 1978; Nicolis and Prigogine, 1989; Aon and Cortassa, 1997).

It is argued, at the state of far-from-equilibrium when the constraints from the host environment are sufficiently strong, the open system can adjust to its host environment in several different ways to the emergence of order in the system (Nicolis and Prigogine, 1989). In other words, during the far-from-equilibrium state, the open system is highly sensitive to external influences; any small change in the host environment will trigger fluctuations, which is remarkable space-time self-organisations of energy and matter distributions in the system, to reduce entropy into the host environment (Nicolis and Prigogine, 1977; Ferracin, Panichelli, et al, 1978; Sahal, 1979; Kay, 1984; Prigogine and Stengers, 1984). Internally, a self-organising structure arises in open systems whose function is to enhance the flux of inputs from the host environment and to optimise energy and matter distribution (Nicolis and Prigogine, 1989; Schneider and Sagan, 2005).

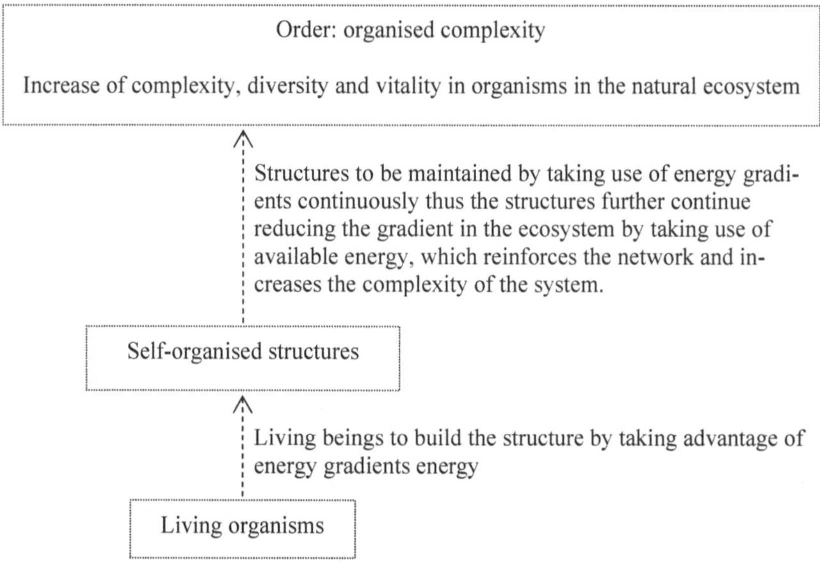

Figure 3–6 Living organisms' structure for self-organisation in open systems evolution

In brief, it is the state of non-linear far-from-equilibrium during evolutionary thermodynamics that provides the open system sufficient randomness so that the difference between the past and the future irreversibly enters its description (Prigogine and Stengers, 1984). All of biology lies within the near-equilibrium region of thermodynamics. A specific material form at far-from-equilibrium is a complex organic system of interconnected cycles, in which living organisms produce structures by using available energy, in order to maintained or suspend metabolism (Schneider and Sagan, 2005), as shown in Fig 3–6. A prototype of self-organisation phenomena in nature is thermal convection (Nicolis and Prigogine, 1989), as show in Fig 3–7.

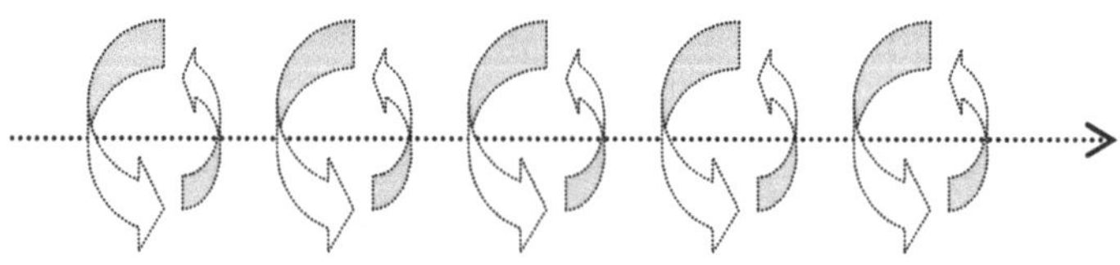

Figure 3–7 Diagram of the spiral as a structural element of patterns in nature

3.1.2.4 Thermodynamic non-equilibrium in open systems

The completion of open systems evolution refers to the state of thermodynamic non-equilibrium when the variations of the open system are constant, e. g. the rate of entropy is zero and the entropy production is minimised. At this state, the open system is highly organised regarding to the material and energy distribution, the emergence of order. After the completion of evolution, the open system will maintain the stable and compatible thermodynamic exchange with the host environment.

The thermodynamic non-equilibrium state of open systems is different from thermodynamic equilibrium state of isolated and closed systems (Prigogine, 1961; Nicolis and Prigogine, 1989; Aon and Cortassa, 1997; Sandler, 2006). The thermodynamic equilibrium state of isolated and closed systems is a state when thermal differences and gradients are zero, entropy production is maximized, and no more chance for any thermodynamic transformation. In contrast, an open system at the steady state of non-equilibrium is maintained in distance from equilibrium, since there are continuous fluxes of inputs from the host environment, thus the system is dissipating inputs and producing entropy at the minimal value.

In brief, an open system as dissipative structure is a regime charactered of symmetry breaking in multiple choices and correlations of a macroscopic range for the birth of organized complexity. In nature, non-living organised systems such as tornadoes and lasers, as well as living systems from cells to ecosystems, obey the rules of non-equilibrium thermodynamics in a nonlinear feedback loop (Prigogine and Stengers, 1984; Nicolis and Prigogine, 1989).

3.1.2.5 Minimal entropy production

In physics, the steady-state or stationary state of non-equilibrium in open thermodynamic systems is also called entropy balance state, which means, through the evolutionary thermodynamics, the value of entropy decrease from the initial value to a minimal value, compatible with the constraints imposed by the host environment (Prigogine, 1961; 1978; Aon and Cortassa, 1997; Prigogine and Stengers, 1984). At this state, the rate of positive entropy production from the open system is compensated by the rate of negative entropy of matter, energy and information from the host environment in such a way that the total rate of entropy is zero. Thus, the value of entropy can work as a quantitative indicator of transitional phases of open systems' evolutionary thermodynamics (Prigogine and Stengers, 1984). It informs the on-going microscopic thermodynamic dispersal configuration of energy, matter and information over the evolution of open system.

3.1.2.6 Emergence of order: organized complexity

In addition to minimal entropy production at the completion of evolutionary thermodynamics, order or organized complexity emerges at the macroscopic scale of the open system. It refers to a highly organized structure in the open system through self-organization of the open system in the context of the host environment (Prigogine and Stengers, 1984; Nicolis and Prigogine, 1989; Crutchfield, 1994; Epstein and Axtell, 1996; Schneider and Sagan, 2005; Prokopenko, Boschetti and Ryan, 2007).

3.1.3 Summary

In summary, the Second Law of Thermodynamics states that thermodynamic transformations of isolated and closed systems towards disorder at thermodynamic equilibrium, and of open systems towards order or organised complexity at thermodynamic non-equilibrium (Prigogine and Stengers, 1984). The general mechanisms of open systems thermodynamics is systems capturing the existence of gradients and degrading of inputs of energy, matter and information from the host environment, converting some inputs for effective work to upgrade the system's organisation towards order, with some irreversibly unavailable and wasted energy as entropy. Hence, there are two parallel during the evolutionary thermodynamics of open systems: the first one is the macroscopically observable entropy bring up an irreversible negative impact, normally called "the arrow of time"; another is the microscopically self-organised complexity as order of the open system, bring up a positive impact, which can be called as "the second arrow of time".

Applying cosmology, the first arrow of time refers to entropy increases irreversibly; the second is the cosmic evolution of organic lives surviving and evolving to highly organized complexity on the earth through a steady growth of structure supported by the solar energy in the universe. Such an axiology revealed by the Second Law of Thermodynamics dramatizes what amounts to a cosmic battle between evil and good: the forces of entropy and the forces of increasing complexity.

3.2 The concept of entropy

To quantitatively understand these statements of the Second Law of Thermodynamics, mathematical expressions are necessary. Central to the Second Law of Thermodynamics, entropy is such a mathematical term describing thermodynamics. Originally, entropy is used for macroscopic descriptions of thermodynamic transfers in isolated, closed and open systems, such as an internal property of matter expressing heat and temperature ratios of the status quo of matter, the spontaneous thermodynamic change or dissipation of energy to degrade the existence of gradients, and the irreversible tendency of increasing wasted and unavailable energy during thermodynamic transformations of isolated, closed and open systems. In statistical mechanics, entropy has been extended to a microscopic description of thermodynamic systems corresponding to macroscopic thermodynamic states, i.e. disorder, randomness and other probability arrangements of particles at molecular, atomic levels. Based on these statistical mechanics meanings, entropy refers to the uncertainty of information transfer in information science and the uncertainty of measurement in quantum theory.

3.2.1 *Entropy in classical thermodynamics*

3.2.1.1 *Entropy representing the status quo of thermodynamic systems*

The original meaning of entropy is to represent macroscopic thermodynamic properties and functions of thermodynamic systems in classical thermodynamics (Clausius, 1850; Atkins, 1984). The term of entropy was first noticed in thermal machines by Carnot and Clausius for the initial statement of the Second Law of Thermodynamics, and lately studied in gas and kinetic theory by Ludwig Boltzmann, Josiah W. Gibbs and others (Nicolis and Prigogine, 1989). Before the formal establishment of the Second Law of Thermodynamics, a theorem respecting the equivalence-values of the transformations was developed by Clausius (1850; 1886). It states the algebraic sum of all the transformations occurring in a cyclical process can only be positive, or, as an extreme case of reversible cyclical process, can equal to zero. This theorem is expressed in the following equation (3–1):

$$\oint \frac{\delta q}{T} \geq 0 \qquad (3\text{–}1)$$

where

- δq is an element of the heat given up by the system to any reservoir of heat during the changes, heat which it may absorb from a reservoir being reckoned as negative; for every cyclical process, including irreversible and reversible processes, equality applies only to reversible cyclical process (Prigogine, 1961);
- T is the absolute temperature of the body at the moment of giving up this heat.

Based on it, the Second Law of Thermodynamics was first formulated (Clausius, 1865; Prigogine, 1961), which states, during thermodynamic transformations in isolated systems, a function of the state varies monotonically positive and tends to maximum at the final state of thermodynamic equilibrium. This state function of thermodynamic systems, a fundamental physical property depending only on the current state of the system (Schrödinger, 1962), is defined as entropy, symbolized as S. Quantitatively entropy is measured in thermodynamic experiments as energy divided by temperature, the ratio of heat to temperature as shown in Equation (3–2). It implies entropy will increase when heat passes into a system (for Q_{rev} is then positive); entropy will decrease when heat passes out of a system (for Q_{rev} is then negative); the total entropy change around the closed circuit of the Carnot cycle, $\Delta S = \sum \frac{Q_{rev}}{T} = 0$ (Roy, 1995).

$$\Delta S = \frac{\delta q_{rev}}{T} \tag{3–2}$$

where

- δq_{rev} is the sum of the incremental values, and constitutes each element's or compound's standard molar entropy, i.e. the reversible thermal change of heat;
- T is a positive quantity of absolute temperature;
- ΔS is entropy change, a fundamental physical property and an indicator of the amount of energy stored by a substance at absolute temperature.

A general correlation between temperature and molecular entropy can be demonstrated in the heat or temperature curve changing in water (Stonier, 1990; Moore, Stanitski, et al., 2001; Jungermann, 2006; Atkins and Paula, 2006), as shown in Fig 3–8, in which lower temperature corresponds to smaller entropy, while higher temperature corresponds to larger entropy. It suggests at the absolute zero point of temperature, -273 °C, the entropy of any substance is zero, which is the manifestation of the Third Law of Thermodynamics, the unattainability of absolute zero (Schrödinger, 1962). When a mole of substance at 0 K is warmed by its surroundings to 273 K, the sum of the incremental values of $\delta q_{rev}/T$ constitute each compound's standard molar entropy, a fundamental physical property and an indicator of the amount of energy stored by a substance at 273 K (Clausius, 1865; Sandler, 2006).

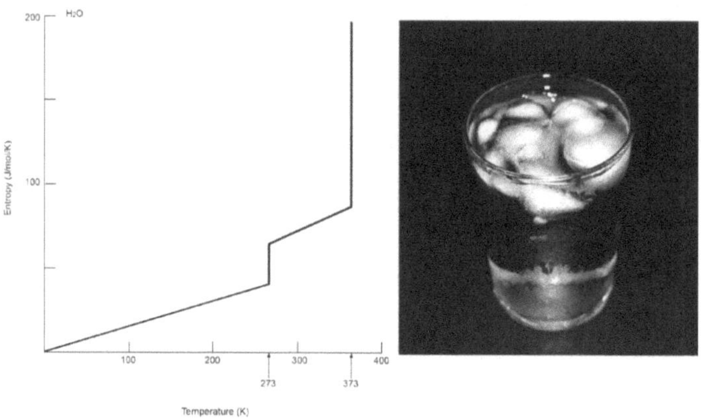

Figure 3–8 Molecular entropy increases with the temperature of water[24]

When the state of crystalline solid, an example of the equilibrium of optimal organisation, is melt from liquid to gaseous state, from 273K to 373 K, molecular entropy increases by the amount of the heat of fusion divided by the temperature (Schrödinger, 1962; Prigogine and Stengers, 1984). After the point of 373K, molecular entropy increases in an incalculable way, which suggests the structure of water is drastically changing, where the previous linear interrelation of molecular entropy and temperature of water no longer exists. In brief, the change of entropy as a state function of a thermodynamic system indicates a change in the system's structure, due to the heat received by the system, which is applicable not only to changes in inter-molecular organisation, but also to sub-molecular, sub-atomic, and sub-nuclear organisation (Odum, 1996), as shown in Fig. 3–8.

24 Diagram source: Stonier, T. (1990) Information and the internal structure of the universe: an exploration into information physics. London; New York: Springer-Verlag, Copyright permission of diagram is granted for the reproduction in this book by the publisher. Photography of a glass of ice-melt is taken by the author.

3.2.1.2 Entropy representing thermodynamic transformations

In addition to the representation of a thermodynamic system's status quo, in classical thermodynamics, entropy also indicates changes in the status quo of a thermodynamic system, which means a state function representing the transition of a system from one state to another (Haddad, Chellaboina, and Nersesov, 2005). For example, entropy refers to the irreversible tendency to increased disorder of isolated systems. Based on this meaning, entropy in biological thermodynamics is an index of a system's tendency to spontaneous changes (Haynie, 2001); and in physical chemistry, entropy is a technical term to measure energy dispersal at a specific temperature (Atkins and Paula, 2006). This interpretation of entropy can be found in the original meaning of Greek words "τροπη" and "ευτροπη", which means transformation, evolution, mutation or even confusion (Prigogine, 1961; Huber and Mills, 2006).

3.2.1.3 Entropy representing thermodynamic degradation and unavailability

Over the thermodynamic transformations in isolated, closed and open systems, the amount of energy unavailable to do work is defined as entropy (Clausius, 1865; Pearsall and Hanks, 1998; Pearsall and Trumble, 2003; Daintith, 2005; Carnot, 2005; Haddad and Chellaboina, et al. 2005). In physics, entropy is defined as a measure of the irreversible partial loss of the ability of a system to perform work (Cutnell and Johnson, 2001). It is also interpreted as the amount of wasted energy in a thermodynamic transformation of a system (Haddad, Chellaboina, and Nersesov, 2005). In biochemistry, entropy is defined as energy broken down in irretrievable heat (Rosnay, 1979). The degradation of energy in macroscopic level can be explained from a microscopic perspective in chemistry (McGraw-Hill concise encyclopaedia of chemistry, 2004) where entropy is defined as a non-conserved thermodynamic state function, measured in terms of the number of microstates corresponding to macroscopic degradation in usable energy, which will be specified in statistical mechanics in the following.

This meaning of entropy, unavailable and unusable energy given up to the system's surrounding during spontaneous thermodynamic changes due to gradients, is expressed in Gibbs free energy equation, as shown in Equation (3–3).

$$\Delta G = \Delta H - T\Delta S \tag{3–3}$$

where

- ΔG is the Gibbs free energy change of the system, the maximum amount of work obtained from the system under any given set of conditions, e.g. constant pressure and temperature;
- ΔH is the enthalpy change, the total energy given up by a system;
- T is the absolute temperature;
- ΔS is the entropy change;
- $T\Delta S$ is the energy given up to the system's surrounding as unusable heat.

This equation describes the direction and quantifies the magnitude of the spontaneity of heat transfer from one system to another (Prigogine and Stengers, 1984). It quantifies if the gradients between the thermodynamic system and another system or the host environment is sufficient for thermodynamic transformation, as the gradients determine the spontaneity of thermodynamic transformation. In terms of this equation, a thermodynamic exchange can be predicted staying away from or spontaneously going to thermodynamic equilibrium. For example, a system is maintained at thermodynamic equilibrium when the Gibbs free energy is at a minimum, entropy is at a maximum, and no more thermodynamic exchanges.

As introduced above, a complete configuration for thermodynamic transformations consists of two systems in gradients, or a thermodynamic system and the host environment. While heat transfers from a hotter system to a colder system or from the system to the host environment, entropy irreversibly changes in both parties since their status quo of heat and temperature will change and tend to equal towards thermodynamic equilibrium. Thus, entropy is produced by both parties of the system and the surrounding or both of the systems, as shown in Equation (3–4).

$$\Delta S_{universe} = \Delta S_{surroundings} + \Delta S_{system} \quad (3\text{–}4)$$

where

- $\Delta S_{universe}$ is entropy produced by the whole universe for thermodynamic transformation;
- ΔS_{system} is entropy produced by the system; and
- $\Delta S_{surroundings}$ is entropy produced by the surroundings, e.g. the host environment or another thermodynamic system.

3.2.1.4 *Entropy representing entropy balance state of open systems*

It is argued, the thermodynamics of irreversible processes is entirely concerned with the analysis of entropy production and the study of the relations between rates and affinities (Jaynes, 1957; Prigogine, 1961). To describe irreversible thermodynamics, the classical statement represented by the value change of variables could not disclose the dynamics specifically as the differential equation does; while the rates, or variations with time, are the effective terms to describe the evolutionary thermodynamics of open systems (Prigogine, 1961; Aon and Cortassa, 1997; Sandler, 2006). Represented in the rates of energy, matter fluxes across the system boundary, entropy balance for open thermodynamic system expresses the change of entropy at the steady-state continuous operation (Sandler, 1997; 2006), as shown in Equation (3–5).

$$\frac{dS}{dt} = \sum_{k-1}^{K} \dot{M}_k \hat{S}_k + \frac{\dot{Q}}{T} + \dot{S}_{gen} \quad (3\text{–}5)$$

where

- dS/dt is the rate of entropy production of an open thermodynamic system;
- $\sum_{k=1}^{K} \dot{M}_k \hat{S}_k$ is the net rate of entropy flow due to the flows of mass into and out of the system, $\hat{S}_k$ = entropy per unit mass, $\dot{M}_k$ = the rate of net matter flow across the boundary of the system;
- $\dot{Q}/T$ is the rate of entropy flow due to the flow of heat across the system boundary;
- $\dot{S}_{gen}$ is the rate of internal generation of entropy within the system; $\dot{S}_{gen} > 0$ for irreversible thermodynamic processes, it means the inter-

nally generated entropy is positive, $\dot{S}_{gen} = 0$ only for reversible process.

- With the mathematical index of entropy, a comparison of isolated systems and open systems is generalised in Fig 3–9.

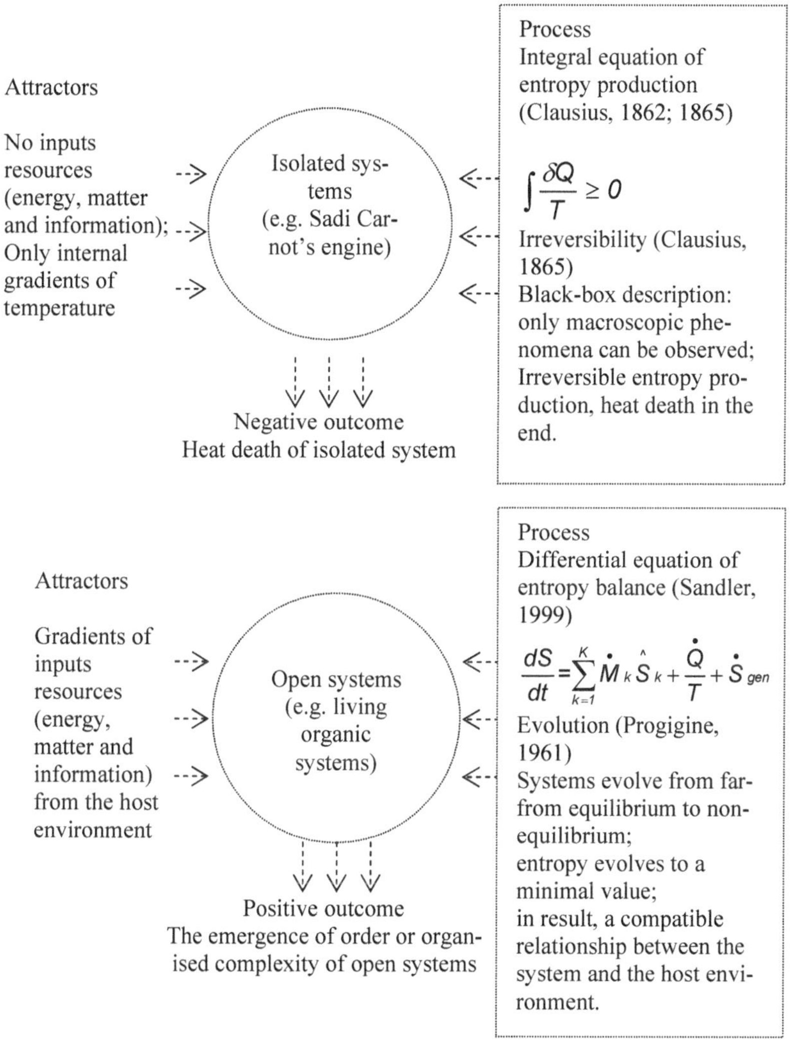

Figure 3–9 A comparison of open systems and isolated systems indexed by entropy

3.2.1.5 *Entropy representing phase transitions of evolutionary thermodynamics*

According to entropy balance of open thermodynamic system in Equation (3–5), it can be generalised, there are two variations of entropy in open systems (Prigogine and Stengers, 1984; Morin, 1995). One is the exchanges with the external environment in neg-entropy of matter and energy from the host environment across the boundary which produces the external entropy $dS_{external}$; another is the interaction among the constituents of the system while the open systems process in self-organization for the establishment of structure and complexity which produces the internal entropy $dS_{internal}$. These two entropy rates are generalised in Equations (3–6), (3–7) and (3–8).

$$\frac{dS}{dt} = \frac{dS_{external}}{dt} + \frac{dS_{internal}}{dt} = \frac{dS_{input}}{dt} + \frac{dS_{self\text{-}gen}}{dt} \qquad (3\text{–}6)$$

$$\frac{dS_{external}}{dt} = \frac{dS_{input}}{dt} = \sum_{k=1}^{K} \dot{M}_k \hat{S}_k + \frac{\dot{Q}}{T} \qquad (3\text{–}7)$$

$$\frac{dS_{internal}}{dt} = \frac{dS_{self-gen}}{dt} = \dot{S}_{self-gen} \qquad (3\text{–}8)$$

where

- dS / dt is the total rate of entropy;
- $dS_{external} / dt$ is the rate of entropy of exchange processes, e.g. energy and matter flows from its host environment; and
- $dS_{internal} / dt = \dot{S}_{self-gen}$ is the rate of entropy production by the irreversible processes within the open system, such as chemical reactions, heat conduction, diffusion, viscous dissipation, relaxation phenomena in electrically or magnetically polarized systems, etc, usually it is assumed (Prigogine, 1978) $S_{internal} > 0$.

Thus, the steady state or non-equilibrium of evolutionary thermodynamics of open systems, when the rate of entropy is zero, can be expressed as Equation (3–9):

$$\frac{dS}{dt} = \frac{dS_{int\,ernal}}{dt} + \frac{dS_{external}}{dt} = 0 \quad or \quad \frac{dS_{external}}{dt} = \frac{-dS_{int\,ernal}}{dt} \tag{3–9}$$

where $dS_{int\,ernal} > 0,\ dS_{external} < 0$.

This entropy balance equation implies, the rate of positive entropy produced by the open system, which has negative impact to produce disorder in the host environment, is compensated by the rate of neg-entropy flow, e.g. energy and matter flow from the host environment, which has positive impact to provide the order in the open system (Prigogine and Stengers, 1984; Nicolis and Prigogine, 1989). Thus, by feeding up the neg-entropy from the host environment, the negative impact of open systems to the host environment, positive entropy, is minimised at the non-equilibrium state when the rate of entropy is zero. Furthermore, in biological thermodynamics applicable to the open systems of biological organisms which have the capability to absorb sufficient energy and matter flux to compensate the positive entropy production after the state of entropy balance, for their highly organised complexity, evolutionary thermodynamics can result in $dS_{external} > -dS_{int\,ernal}$, when the rate of entropy production induced by energy and mass fluxes from the host environment is larger than the internal entropy production (Schrödinger, 1962; Bertalanffy, 1968; Aon and Cortassa, 1997). In brief, with the quantitative index of entropy, the transitional phases of open systems evolution is generalised in Tab 3–3. Mathematical descriptions of open systems evolution as dissipative structure indicated by entropy is demonstrated in Appendix: a diagram of transitional phases of open systems evolution.

Table 3–3 Transitional phases of open systems evolution indicated by entropy

States	$dS_{[illegible]}$ & $dS_{[illegible]}$	dS	S	Impact
Far-from-equilibrium	$dS_{external} < -dS_{int\,ernal}$	$dS > 0$	$S_{initial}$	Negative (disorder)
Non-equilibrium	$dS_{external} = -dS_{int\,ernal}$	$dS = 0$	$S_{[illegible]} = S_{[illegible]}$	Neutral (balance)
Order emergence	$dS_{external} > -dS_{int\,ernal}$	$dS < 0$	N/A	Positive (order)

Notes	$\frac{dS}{dt} = \frac{dS_{external}}{dt} + \frac{dS_{internal}}{dt}$ $\frac{dS_{external}}{dt} = \sum_{k=1}^{K} \dot{M}_k \hat{S}_k + \frac{\dot{Q}}{T}$, $\frac{dS_{internal}}{dt} = \dot{S}_{self-gen}$

However, the total entropy over the time course of open systems evolution increases, represented in Equation (3–10):

$$S = \sum_{t=initial}^{t=non\text{-}equilibrium} S(t) \qquad (3\text{–}10)$$

where

- S means the total entropy is accumulated by entropy produced over the time course of evolution, from initial maximum value $S_{initial}$ to final minimal value $S_{minimal}$;
- $S_{initial}$ is the initial maximum entropy value, implying the maximum disorder;
- $S_{minimal}$ is attained when the rate of entropy is zero at the non-equilibrium state.

3.2.1.6 Entropy produced by physical and chemical potentials

At the microscopic level, entropy production is explained as physical and chemical potential in the spontaneous change within the system to attain the physicochemical thermodynamic equilibrium (Prigogine, 1961; Prigogine and Stengers, 1984), as shown in Equation (3–11). This equation suggests entropy production results from a transport of heat and matter between the two phases of the system and also from the chemical reaction taking place in each phase.

$$dS = \frac{dE}{T} + \frac{dV * p}{T} - \sum \left(\frac{\mu_Y}{T}\right) * dn_Y \qquad (3\text{–}11)$$

where

- E and V are the variables of the function of thermodynamics in the system;
- The quantities μ_Y are chemical potentials in the thermodynamics of physicochemical equilibrium.

3.2.2 Entropy in statistical mechanics and information science

Based on the original meanings in classical thermodynamics, entropy is used in statistical mechanics, probability theory and information science to represent microscopic configuration at molecular and atomic scales (Prigogine and Stengers, 1984; Roy, 1995; Li and Vitányi, 2001; Schneider and Sagan, 2005; Huber and Mills, 2006; Stowe, 2007). In statistic mechanics, for example, entropy measures disorder, randomness and uncertain at the microscopic scale, corresponding to unavailability and degradation at the macroscopic scale of thermodynamic systems. Thus, microscopic and macroscopic behaviours of thermodynamics systems can be configured in integration by both statistical mechanics and classical thermodynamics, in which, entropy represents both thermodynamics and information senses (Shannon, 1949; Schneider and Sagan, 2005).

3.2.2.1 Boltzmann's entropy: microscopic probability of thermodynamic systems

In gas and kinetic theories, the basis of thermodynamic statistics, the increase of entropy with the increase in temperature is interpreted related to the increasingly random motion of molecules in the system (Boltzmann, 1964; Prigogine and Stengers, 1984; Schneider and Sagan, 2005). This correlation of the system's heat content and the system's organisation implies probability function of entropy in statistics, which recasts thermodynamics as the microscopic statistical distribution of energy in a system, while entropy refers to the probability of particular microscopic arrangements of the system (Schroeder, 1999; Pearsall and Trumble, 2003).

Precisely, an essential equation of entropy in statistical mechanics represents the distribution of a given amount of energy E over N identical systems, where entropy is measured as Boltzmann's constant times the logarithm of a multiplicity (Schrödinger, 1962; Boltzmann, 1964; Stonier, 1990; Baierlein, 1999), as shown in Equation (3–12). For example, for a given set of macroscopic variables, like temperature and volume, entropy measures the degree to which the probability of the system is spread out over different possible states; the more states available to the system with higher probability, the greater the entropy. In brief, the multiplicity of a macro-state is the number of micro-states which corresponds to the macro-state (Baierlein, 1999).

$$S = k_B \ln \Omega \tag{3–12}$$

where

- k_B is Boltzmann's constant $1.38066 \times 10^{-23} J\,K^{-1}$;
- Ω is the number of microstates corresponding to the observed thermodynamic macro-state, the probability of a system being in a particular microstate, given that it is in a particular macro-state;
- S is the logarithm of probability

3.2.2.2 Disorder and randomness of a system

Based on the probability meaning of microscopic arrangement in a system, entropy in statistical mechanism is further defined as a quantitative measure of disorder, which is a system function of disorganisation, the increasingly randomness of the constituents in a system, particularly at the atomic, ionic or molecular level (Schrödinger, 1962; Brooks and Wiley, et al., 1988; Nicolis and Prigogine, 1989; Pearsall and Hanks, 1998; Roy, 1995; Callen, 1985; Moore, Stanitski and Jurs, 2001; Clark, 2004; Haddad, Chellaboina and Nersesov, 2005; Daintith, 2005; Chang, 2006). An increase in entropy represents a loss of organisation, more disorder and more probability, expressed in Boltzmann and Schrödinger Equation, as shown in Equation (3–13).

$$S = k \log D \tag{3–13}$$

where

- K is Boltzmann's constant 3.2983×10^{-24} *cal/deg* ;
- D is a quantitative measure of the atomistic disorder of the body in question (Stonier, 1990).

The loss of order is equivalent to a rise in chaos, thus entropy refers to chaos (Schneider and Sagan, 2005; Moore, 2006). The relative representations of disorder and order of a system are generalised (Brooks and Wiley, 1988; Callen, 1985; Haddad, Chellaboina and Nersesov, 2005) in the following equation (3–14):

$$Disorder = C_D/C_I \qquad Order = 1 - C_O/C_I \tag{3–14}$$

where

- C_D is the "disorder" capacity of the system, which is the entropy of the parts contained in the permitted ensemble;
- C_I is the information capacity of the system, an expression similar to Shannon's channel capacity; and
- C_O is the "order" capacity of the system.

3.2.2.3 Maximum entropy principle for an optimal distribution

In terms of statistical mechanics, the irreversible entropy production in isolated, closed systems and open systems can be interpreted in probability terms (Schneider and Sagan, 2005). In isolated systems, the irreversible entropy increases to maximum value at thermodynamic equilibrium, which is generalised as the most probable state, termed as maximum entropy probability (Jaynes, 1957; Stowe, 2007). In evolutionary thermodynamics of open systems, at the initial stage, a high value of entropy implies a highly disordered state of a system associated with a high probability, which corresponds to the state of maximal disorder (Nicolis and Prigogine, 1989). The final non-equilibrium state with minimum entropy production implies an order state of a system associated with low entropy. This highly organised state of open thermodynamic systems is a state of high statistical improbability (Bertalanffy, 1968). However, the total entropy production into the host environment reaches maximum, which has been accumulated over the evolution. This trendy of open systems' evolution finishing at non-equilibrium state is generalised as maximum entropy principle in probability theory (Jaynes, 2003), represented in Equation (3–15).

$$H(p_1 \dots p_n) = -K \sum_{i=1}^{n} p_i \ln p_i, \sum_{i=1}^{n} p_i = 1 \tag{3–15}$$

where

- H is information entropy for equal probabilities p_i;
- K is Boltzmann's constant which determines the units of entropy.

In probability theory, a fair allocation refers to the specific distribution when the proportion of resource to each unit is allocated exactly as how much it deserves, which is termed as canonical distribution. The principle of maximum entropy has the clearest theoretical justification as an optimal distribution, which is the most highly developed computationally non-speculative procedure to yield the optimal inference from the incomplete information (Jaynes, 2003). Maximum entropy is the least biased estimate possible on the given information, i.e. maximally noncommittal with regarding to missing information, and a mathematical procedure for the most unbiased representation of the knowledge of the state of a system. In brief, maximum entropy principle is an approach to attain practical certainty under uncertainty. It implies no possibility is ignored; it assigns positive weight to every situation that is not absolutely excluded by the given information. Thus, maximum entropy principle is argued the optimum distribution, the most frequency distribution in the natural world, as a broadest distribution represents more uncertainty than does a sharply peaked one.

Maximum entropy principle in information science is also applicable in sociological science (Hagan, 2001). For example, the ethic meaning of maximum entropy probability distribution which means the most possibly broadest access of resource to the agents within the system, with reference to the constraints in host environment. In the real world, 20% of population of the world consume 80% of resource in the world (2007 Human Development Report), which is certainly not an optimal distribution at non-equilibrium or equilibrium in terms of open system evolution. It requires a further global redistribution in maximum entropy principle for equality and sustainability. To realise such a global coordinated optimal distribution, information of each unit in the global system and information of available resources are crucial.

3.2.3 Shannon entropy in information communication theory

In information communication theory, information is associated with the amount of freedom of choice in constructing messages, which represents the probability and the diversity of the system, rather than representing how organised the system is. Based on the concept of uncertainty and disorder in statistic mechanics, Shannon and Weaver (1949) developed a meaning of entropy in information communication theory, which is usually called Shannon entropy (Shannon, 1949; Schrödinger, 1962; Stonier, 1990; Baierlein, 1999; Balian, 2004). It describes the uncertainties associated with the utilisation of characters in sending and receiving messages, i.e. a function representing the amount of information which is missing before reception, as shown in Fig 3–11.

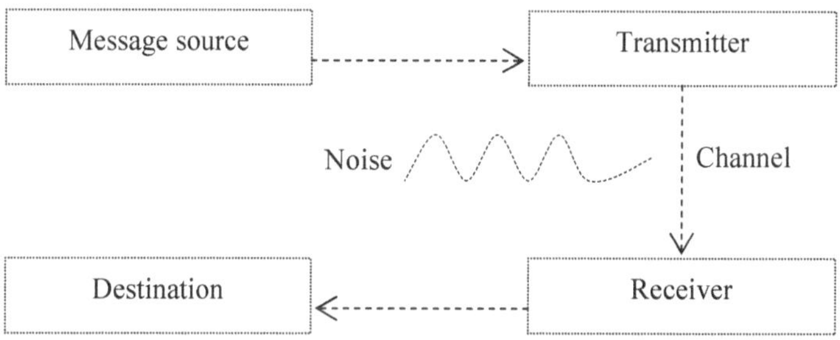

Figure 3–10 Shannon's model of a communication process

Symbolised as H, mathematically, Shannon's entropy is a logarithmic measure of the rate of transfer of information in a message or in particular language, referring to missing information during communicating a message between a sender X and a receiver Y, i.e. the number of digits in probability theory (Shannon, 1949; Stonier, 1990; Prokopenko, Boschetti, and Ryan, 2007), as shown in Equation (3–16).

$$H = K\,ln(\frac{1}{p}) \qquad (3\text{–}16)$$

where

– H is information entropy for equal probabilities $p_i = p$;

– K is a constant which determines the units of entropy, for example, if the units are bits, then $K = \frac{1}{ln(2)}$.

3.2.3.1 *Entropy in information science: uncertainty of information distribution*

Extension to Shannon's entropy, entropy measures the amount of uncertainty and randomness in information science, representing probability distribution of information, in Equation (3–17).

$$H(X) = -\sum_{x \in X} P(x) log\, P(x) \qquad (3–17)$$

where

– K is a constant which determines the units of entropy, if the units are bits, then $K = \frac{1}{ln(2)}$;
– H is information entropy for equal probabilities p.

Both Boltzmann's entropy and Shannon's entropy shows increasing in time until a certain state is reached, such as maximum entropy production of isolated systems' equilibrium. The intrinsic dynamics of information, indicated by the dynamics of entropy, implies the inner structure of information (Li and Vitányi, 2001; Prokopenko, Boschetti and Ryan, 2007). In addition, entropy proliferates in dynamic systems theory, such as metric entropy, topological entropy, algorithmic entropy, in which entropy is the fractal dimension of an appropriate compact set. These various mathematical subspecies of informational entropy share a concern with unpredictability, incompressibility, asymmetry, or delayed recurrence.

3.2.3.2 *Entropy in evolutionary biology: uncertainty of the evolution*

A model of information systems composed by structure and dynamical evolution can be found in evolutionary biology which studies the evolution of biological systems, represented in physical complexity computing information transformation between a biological individual entity and its niche of natural ecosystem over its evolution (Li and Vitányi, 2001; Adami, 2002; Prokopenko, Boschetti, and Ryan, 2007).

Precisely, physical complexity is defined as the amount of information that a population X stores about the environment, niche, E in which it evolves, which is represented as $I(X:E)$. During the increase of the complexity, i.e. the information communication between the biological systems and its niche is measured by entropy, symbolised as $H(X)$, which is a measure of uncertainty (Adami, 2002), formulated in Equation (3–18):

$$H(X) = -\sum_{i=1}^{n} p_i \log p_i, \sum_{i=1}^{n} p_i = 1 \tag{3–18}$$

where

- $H(X)$ is entropy of an ensemble (i.e., a population) of sequences X;
- p_i is the probabilities of sequences s_i, in the sequence of X; it is a notion of conditional probability, with the reference to its host environment E, niche.

The difference of H_{max} maximum information of the individual biological system about the host environment, and $H(X|E)$ conditional entropy, the potential knowledge (Adami, 2002) possible to evolve as physical complexity, is formulated in Equation (3–19):

$$I(X:E) = H_{max} - H(X|E) = L + \sum_{i=1}^{n} p_i \log p_i, \sum_{i=1}^{n} p_i = 1 \tag{3–19}$$

where

- $I(X:E)$ is entropy shared between X and E, i.e., the information that X has about E;
- $H(X|E)$ is the conditional entropy of X given the environment of E;
- H_{max} is maximum entropy, $H_{max} = L$, the length of the sequence, when all sequences are equally probable in ensemble X, under the condition of no selection occurrence.
- p_i is the probabilities of sequences s_i, in the sequence of X; it is a notion of conditional probability, with reference to its host environment E, niche.

Thus, the evolution of complexity in evolutionary biology is driven by the evolution of entropy, i.e. entropy as potential knowledge to evolve as physical complexity of the biological individual system, through the interactive information communication between itself and its niche during the biological system's evolution and adaptation (Adami, 2002), as displayed in Table 3–4. In other words, during the adaptation of the biological systems embedded in the niche of natural ecosystem, it is the release of entropy as uncertainty which brings up the generation of physical complexity in the biological system. Thus, the more entropy is released, the more physical complexity the biological system increases.

Table 3–4 The evolutions of entropy and complexity in evolutionary biology

<table>
<tr><td rowspan="4">Evolutionary biology</td><td colspan="2">Empty tape: $H_{max} = L$ maximum entropy of a sequence</td></tr>
<tr><td>Information of a sequence $I(X : E)$</td><td>Entropy: $H(X|E) = -\sum_{i=1}^{n} p_i \log p_i$, uncertainty, potential knowledge</td></tr>
<tr><td>Information of a sequence</td><td>Entropy: uncertainty, potential knowledge</td></tr>
<tr><td>Information of a sequence</td><td>Entropy</td></tr>
</table>

In terms of this interpretation of entropy in evolutionary biology, the artistic creation and expression are in fact the reflection of the information stored in living organic human beings as open systems about the natural world as their niche. The more information transferring or attained from the niche of nature, the more complexity the human beings become, the more they become knowledgeable which will benefit their creation in the natural world.

3.2.3.3 Entropy in quantum theory: uncertainty in quantum measurement

Entropy is also applied into quantum mechanics, which is a probability and statistical term referring to the inherent uncertainty of quantum mechanical states. This uncertainty and unavailability in quantum dimensions, where the reality is much close to the nature, can be interpreted as such: any contact would change the object to be observed or measured, even through the medium of matter and wave, light, thus the observer's interference to the system is inevitably changes the accuracy of the observation (Nicolis and Prigogine, 1989; Norris, 2000; Schneider and Sagan, 2005; Stowe, 2007). In brief, it is impossible to predict what an electron would do next inside an atom in quantum domain, because of the uncertainty of measurement. The correct mathematical framework for quantum mechanics is established as a theory of measurement, where the usual notion of wave collapse is described as an irreversible process (von Neumann, 1955; Stotland, Pomeransky, et al. 2004; Stowe, 2007). In conjunction with the 53Hdensity matrix, the classical concept of entropy is extended into the quantum domain in an explicit expression, to distinguish between the minimum uncertainty entropy of pure states and the excess statistical entropy of mixtures.

3.2.4 Summary

In summary, the concept of entropy in statistic mechanics refers to the distinction and relevance of macroscopic performance and microscopic configuration of thermodynamic systems in gas theory, with extensional meaning to disorder and uncertainty of a system, thermodynamic systems or information systems, and biological systems, as shown in Fig 3–11.

Gas theory (Boltzmann, 1906)
Statistical interpretation of the Second Law of Thermodynamics: macroscopic phenomena to be explained by microscopic configuration $S = K_B \ln \Omega$

Information communication theory (Shannon, 1949)
Uncertainty (missing information) in information communication

$$H = K \ln(\frac{1}{p})$$

Evolutionary biology (Adami, 2002)
The evolution of complexity by the release of entropy uncertainty during the adaptation of an evolutionary biology $H(X) = -\sum_{i=1}^{n} p_i \log p_i$

Statistic mechanisms (Jaynes, 1957)
Maximum-entropy probability of a system $H(p_1 \dots p_n) = -K \sum_{i=1}^{n} p_i \ln p_i$

Figure 3–11 Information entropy from gas theory to evolutionary biology

3.3 Universal generality of open systems evolution

Pervasive and fundamental in nature, open systems can be found such as living organisms, ecosystems, convection cells, computers, hurricanes, biospheres, in which the energy flow structures are organized in evolutionary thermodynamics. The negative and irreversible consequences of open systems evolution, entropy, which originally refers to thermodynamic dissipation, degradation and unavailability of energy in thermodynamics science, is extensionally referred to metaphors such as chaos, disorder, uncertainty in information science, and equilibrium in organisms such as death, decay, waste, degradation. In the cosmic axiology, entropy is used as a metaphor referring to the necessary unravelling of time. It can be considered 'bad', 'evil', 'sinful', chaotic in the sense of frustration, destruction, boredom, dissolution, decline, decay, killing and kitsch, mass extinctions, asteroids hitting the earth, the Holocaust, warfare, negative-sum games, poverty (Marx, 1963), pollution, inefficiency, waste (DeMarco and Lister, 1987), global warm and climate change (Moore, 2006).

For example, social entropy (Bailey, 1990) is a measure of social system structure related to social equilibrium, having both theoretical and statistical interpretations, i.e. society as macro-societal variables is measured in terms of how the individual as micro-societal variables functions in society. In energy economics, economic entropy is interpreted as a semi-quantitative measure of the irrevocable dissipation, degradation, waste and inefficiency of natural materials and energy flows into the industrial productions in an energetic pyramid, such as the extraction of raw materials, refining, processing, purifying, transportation and domestic heating (Georgescu-Roegen, 1971; Meyer, 1983; Faber, Niemes, et al. 1987; Burley and Foster, 1994; Huber and Mills, 2006). Thus, the most effective way to reduce entropy and to raise efficiency, control pollution is to improve control of the flow of energy through the systems that consume it. In brief, energy economics is operated in a logic structure to organise, move and increase order of human environment by adding high grade energy at one end, and then discard some fraction of it in the form of low-grade heat at the other (Prigogine and Stengers, 1984; Huber and Mills, 2006).

While the positive aspects of open systems evolution, the emergence of order, can be found in the magnificent origin, formation, growth and evolution of life, through which, human technology, economy and culture are developed as structure for self-organisation, adaptation and optimisation. In general, order can briefly be specified as desirable properties of a system, such as high organisation, intelligence, efficiency, beauty, humanity, democracy, autonomy, simplicity and even complexity. Especially, humanity has been used as political slogan to campaign for an order of the world.

In brief, the Second Law of Thermodynamics for open systems, or the model of open systems evolution, holds a universal generality to formulate and interpret emergence and transformations in both natural sciences and sociological sciences.

3.3.1 Open systems evolution in the real world

3.3.1.1 Creation of the universe in cosmology

In cosmology, the creation of the universe is interpreted as a thermodynamic production following the Second Law of Thermodynamics, in which, the creation of matter acts as releasing the internal energy and transferring energy from gravitation to matter in the cosmological perspective (Prigogine and Stengers, 1984; Nicolis and Prigogine, 1989).

3.3.1.2 Assumption of the model of origin-of-life

In biochemistry, all the existing models of origin-of-life assume that supplies of high-grade energy are synthesized first, mostly by accidents if suitable supplies of energy which is congenial for the occurrence of life are available in the surrounding, and then transported to a gentle environment where they are combined to form organism capable of self-replication (Watson and Crick, 1953; Wicken, 1978; Watson and Crick, 2005; Schneider and Sagan, 2005; Huber and Mills, 2006). The basic thermodynamics of the origin of life requirement goes a two-temperature thermal cycle for energy-capturing, which is sufficient to drive self-replicating chemistry transforming chaos into order.

3.3.1.3 Evolution of living organic lives

In addition to the assumption of origin-of-life, the life of living organisms is also argued as open systems evolution. Living organic systems are open to the host environment of natural ecosystems, dependent upon energy through light or chemical reactions, adaptively interacting with the host environment. In the preliminary essentiality of a logic structure, living organisms self-spontaneously extract and consume high-grade energy and materials from the host environment, coherently converts these ordered flows of physical and chemical energy into effective work, self-organising to increase its complexity in terms of function and interconnection (Watson and Crick, 1953; Prigogine, 1961; Bertalanffy, 1968; Kay, 1984, 2002; Prigogine and Stengers, 1984; Nicolis and Prigogine, 1989; Kauffman, 1993; Schneider and Sagan, 2005). In brief, the characteristics of organic life include metabolism, growth, development, self-reproduction, self-regulation, responses to stimuli, spontaneous activity, and its abilities and potentials of growth and destruction, and perception and self-reflection.

This organised complexity is order of organic systems, which aims to avoid the death of thermodynamic equilibrium and to maintain their survivals from the constraints imposed by the host environment (Schrödinger, 1962; Bertalanffy, 1968; Kay, 1984; Schneider and Sagan, 2005). Meanwhile, living organisms release heat and other thermodynamic wastes as entropy into their host environment (Prigogine, 1961; Schneider and Sagan, 2005). In brief, living organisms attain a stationary and fairly low entropy level for their survival, by continually sucking orderliness from its environment in chemical and physical processes and at the expense of entropy produced in the surrounding. This paradigm of open systems distinguishes from the early paradigm of living machine by Descartesism (Bertalanffy, 1968; Schneider and Sagan, 2005).

In addition to the origin, evolution, and complexity of living organic lives, it is also argued, the complexity of human life, which includes human illness, depression, psychological regression, eating, growing, and healthy and stability of life, depends upon energy transformation (Schneider and Sagan, 2005).

3.3.1.4 Formation and degradation of matter

In material science, the ordered and organized material structures, like crystals and snowflakes in nature, can form spontaneously under the condition by extracting well-ordered power from the poorly ordered environment, if suitable supplies of energy are available for the formation of matter in a congenial place (Prigogine, 1978). In addition, the degradation and fragmentation of material can also be explained as thermodynamic equilibrium in terms of the Second Law of Thermodynamics, in which all points would be uniformly distributed in the accessible region of phase space (Blofeld, 1978; Nicolis and Prigogine, 1989; Walker, 2007). Thus, an entire configuration of thermodynamic capability of materiality can be attributed as growth, decay, exchange, conversion, adaptation, retention, infiltration and evaporation, in terms of thermodynamics laws.

3.3.1.5 Irreversible evolution of natural ecosystems

The regenerative and evolutionary processes of the natural ecosystem, which result in the viability and complexity of the natural ecosystems, is argued as non-equilibrium dissipation and irreversible thermodynamics of energy flows following the Second Law of Thermodynamics (Odum and Odum, 1981; Prigogine and Stengers, 1984; Nicolis and Prigogine, 1989; Schneider and Sagan, 2005). In specific, the ecosystem is defined as a unit of biological organisation, a population of all the organisms in a given area, self-organizing and interacting with the physical surrounding.

3.3.1.6 Open systems evolution of human society

With a view to the history of human societies, one of plausible conceptualisations is that human societies develop as dynamic and complex open systems, which evolve and diffuse outward from simpler forms of organization to more complexity and optimisation, through hunting, agricultural and industrial societies, in analogical mechanisms of order and disorder (Prigogine and Stengers, 1984; Nicolis and Prigogine, 1989; Delanda, 1997; Hagan, 2001; Huber and Mills, 2006; Mainzer, 2008), as generalised in Equation (3–20), and shown in Fig 3–12. Thus, the principle of human development for a long term's sustainability is adaptive and responsive to each dynamic time, stage, situation, an innovative creativity under conditional constraints.

$$\textit{The dynamics} = \textit{human aspirations (e.g. individual projects and desires)} + \textit{the constraints from the environment} + \textit{the differences between human aspirations and the realistic achievement (for the next dynamics)} \quad (3\text{–}20)$$

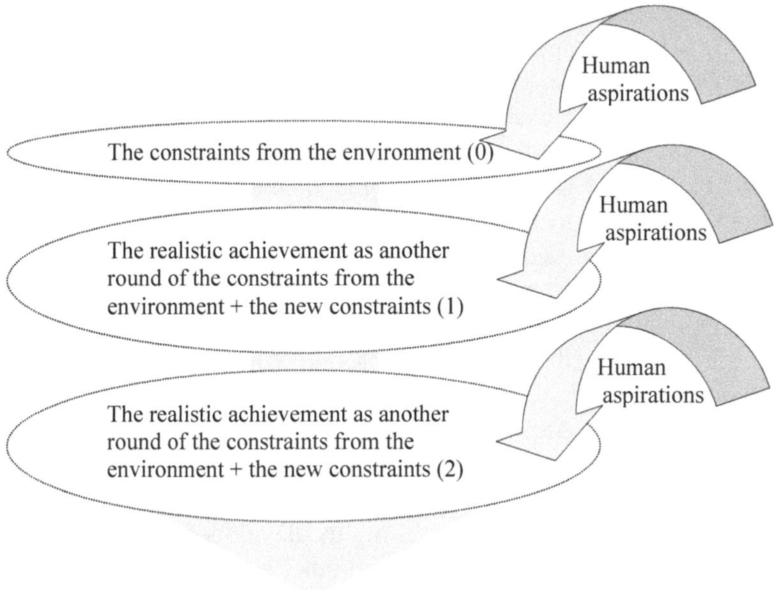

Figure 3–12 Dynamic evolution of human society as an open system

3.3.1.7 Psychodynamics of human psyche as an open or closed system

In psychology, psychologist Jung applied thermodynamics laws (Hall and Nordby, 1973) to interpret the human psyche's growth. Initially, the total personality or psyche is a relatively closed system, energy self-contained apart from any other energy system. It derives its psychic energy primarily from experiences through the sense organs, including touch, see, smell, taste, feel or hear, which contribute as actual or potential external forces to perform psychological work. These stimulations continuously feed the psyche system for its internal redistribution or shifting of energy throughout the structures of the psyche, and the transfer of energy from one structure to another. This dynamics of psyche is termed psychodynamics, which is always in unstable states, never attaining perfect equilibrium. Even the slightest stimulus may have far-reaching consequences on one's mental stability to produce large effects in a person's behaviour.

Two basic principles in terms of the laws of thermodynamics have been derived to analyse the psychodynamics (Hall and Nordby, 1973). The principle of equivalence for the conservation of psychic energy is based on the First Law of Thermodynamic, which states whenever psychic energy is transferred from one element or structure of the psyche to another element or structure, the value of the energy will remain the same. The principle of entropy energy transferring from high value to low value and resulting in equilibrium of psychic forces, based on and the Second Law of Thermodynamics, governs the exchanges of energy throughout the whole personality, to achieve a completely balanced system. In terms of these principles, storms of youth is analysed as the turbulence at young age due to the vast amount of energy that pours into the psyche both from external and bodily sources; and tranquillity of age is the experiences which have been dealt with and incorporated into the personality with some degree of psyche's harmony when both the inner and outer worlds are adapted jointly. Furthermore, a highly developed psyche system always seizes all possible energy from other systems, as an open system automatically capturing psychic energy for its growth, being able to adapt not only to the outer world, but also to the inner psychic world.

In brief, considering psychical brain as an open system which follows the thermodynamics laws, it is generalised, energy use in a physical brain determines the performance of brain, fast, adaptive, intelligent, or slow, inertia, inapt. Thus, the differentials of idiot and genius can be generalised as different thresholds of brain's capacity on implementing energy use, which suggests the possible evolution from idiot to genius by adapting their behaviours according to an alternative formulation of energy usage within the brain and responding intellectual performance of the brain.

3.3.2 Self-organisation of open systems in nature

As generalised above, during the evolution of open systems in both natural world and human evolution, self-organisation is usually processed with some microscopic structure within the open systems, in a logic order to complete the evolution towards an organised complexity.

3.3.2.1 Structure of the universe in information science

In information physics, the universe is interpreted as a physics system combined by energy and organisation (Blofeld, 1978; Stonier, 1990). The surface structure of the universe is constituted by matter and energy which are readily perceivable to human senses, and indexed by entropy; while the internal structure of the universe is organisation indexed in mathematical quantity of information, which is physically encoded as patterns of organisation. Since information system exhibits organisation, and organisation reflects the ordered arrangement of the constituent parts of a system, information and disorder are inversely related, and order is in opposition to disorder. A highly disordered information system is at a high probability state, which is associated with high entropy, whereas an organised, high-information state is associated with low entropy.

According these definitions, an extreme state of entropy in the universe is "zero information and infinite entropy", when the organisation of the system is zero and entropy reaches its highest level, i.e. a disorder state without any structure (Stonier, 1990). Furthermore, the organised complexity of an information system is contributed by the recursive properties of the system as an open system for cyclic changes in the system (Stonier, 1990). The more complex the system, the greater the likelihood of introducing changes into the system for further generation of complexity. This recursive mechanism of information systems is corresponding to the evolutionary attribute of open systems stated in the Second Law of Thermodynamics, the organised complexity of which is generated in an open system evolution.

3.3.2.2 Self-organisation of the universe and the earth

In cosmology science, the entire universe is argued as dissipative structures arising in self-organisation, which is aging and getting better organised, continually advancing to a higher level of complexity as time sweeps by (Jantsch, 1980; Prigogine and Stengers, 1984). Similarly, in earth science, the self-organisation of the earth is interpreted in "Gaia theory", which builds morality and ethics into the universe (Lovelock, 2006). In the early formulation of 'robust Gaia' theory, the biosphere regulates and maintains the climate and atmospheric composition at an optimum for itself. For example, the rainforest is one of the most robust ecosystems. In the recent formulation of "modest Gaia" theory: the earth is seen as a single physiological system, an entity that is alive like other living organisms, i.e. its chemistry and temperature are self-regulated at a state favourable for life. Apparently, the current environmental state of the earth is unfavourable for life. However, the robustness of the earth system is strong enough, not to be killed by homo-sapiens; if homo sapiens does not prove wise enough and is snuffed out by Gaia as she jumps to a new level of self-organisation, life will go on and a new robust history will start, which requires disorganisation and degradation elsewhere (Schneider and Sagan, 2005). In other words, a new level of self-organisation on the earth will start with the price of the old one extinction, being replaced and archived into history.

3.3.2.3 Self-organisation of living organic systems: cellular automata

The term of self-organisation in biological science is used to interpret how living organisms spatial-temporally self-organise non-uniform distribution of multiple biochemical material in chemical bonding at molecular level, resulting in increase in complexity of biological systems to cope with the constraints in the natural environment (Kay, 1984; Prigogine and Stengers, 1984; Casti and Karlqvist, 1986; Kauffman, 1993; Aon and Cortassa, 1997; Kauffman, 2000; Adami, 2001; Epstein, 2006; Prokopenko et al, 2007). After being released across the boundary of an open system, the inputs of energy, matter, and information will be distributed in an evolutionary mechanism towards a non-equilibrium distribution (Eigen, 1971). This complex, spontaneous and adaptive self-organisation is argued more important than natural selection in the evolution of living organic systems (Kauffman, 1993). The complex behaviour of living organisms can be found in cellular automata, the evolutionary mechanisms of which is generalised as genetic or evolutionary algorithm, composed in genomes, chromosomes, cross-over, mutation, or selection, widely applied in biology, mathematics, physics or social systems (von Neumann, 1966; Terzidis, 2008).

3.3.2.4 Self-organization of capitalism system

Capitalism is argued a flexible and adaptive open system, which has survived terrible catastrophes including two world wars and enormous number of smaller wars, and that the end is not yet near (Hagan, 2001). The dynamic evolution of capitalism as an open system can be interpreted as such: the self-interest of individuals at the subject-level and a competitive market at the object-level create an emergent structure at the meta-level towards into the most interest of enormous profit (Jencks, 1997; Sassen, 1994 ; Hagan, 2001), which is driven by the size of population of these systems forward in positive feedback loops; the increase of energy and information, trade of a growing economy may push the system 'far-from-equilibrium' so that the system can spontaneously jump from one level to the next, from early capitalism to mercantile capitalism to monopoly capitalism to welfare-state capitalism. Generalized by Marx (Marx and Engels, 1954–1962), no mode of production disappears until it has developed all the productive forces which it can contain within its social relation of production, as maximum entropy probability principle to determine the emergence of non-equilibrium of an open system, or equilibrium of isolated system, as the final settle-down of self-organization and evolution. At global perspective, the distribution of capitalism has been extremely uneven; some regions have barely reached the stage of early capitalism and some reached late capitalism.

3.3.2.5 Summary: the mechanism of self-organisation of open systems

In summary, self-organisation of open systems means spontaneous response of open systems to the stimulus from the host environment without central director or any explicit instruction template, to maximize useful inputs transformations for optimal distribution of energy and matter within the system and optimal impact of minimal entropy (Kay, 1984; Odum, 1996; Prokopenko, Boschetti and Ryan, 2007). Three distinctive phenomena occur after the release of inputs as follows and shown in Fig 3–13.

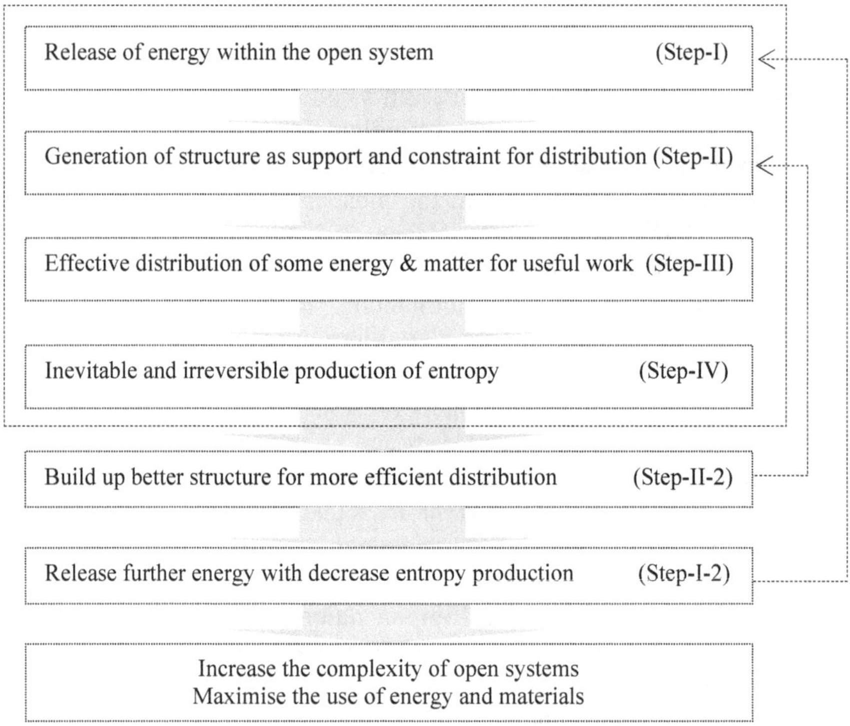

Figure 3–13 Self-organisation for energy and matter distribution in open systems

A structure and some constraints are generated within the system, which are used to control and channel the energy for more efficient release of input(s) further on; most of input resources are effectively distributed within the system for useful work; however, some of energy is inevitably unable for effective work, which is called entropy in classical thermodynamics, which can be considered as an indicator of distribution within the system. The rate of entropy indicates the cycle of distribution of energy and matter within the systems, and the necessity of further structure generation to improve the distribution towards the final optimisation; this optimisation cycle will continue till the rate of entropy is zero and the value of entropy is minimised, when non-equilibrium state is attained by the open system.

3.3.3 Adaptation of open systems in nature

In Darwinism evolution theory of "natural selection on the principle of fitness for survival" and genetic science, adaptation is a powerful mechanism of living systems to increase the probability of the system's fitness for survival; in other words, the success and failure of adaptation will be evaluated to meet the challenges posed by the specific context (Dobzhansky, 1956; Grisogono, 2008). In definition, adaptation refers to how an open system responds to stimulations or pressures from the environment, resulting in the emergence of new properties such as complexity in the system (Prokopenko, Boschetti and Ryan, 2007). Essentially, the mechanism of adaption is composed by three essential functions: generating variety, observing feedback from interactions with the environment, and selection to reinforce some interactions and inhibit others.

These feedback and feed-forward informative interactions between the organism system and the host environment increase the complexity of evolution (Prokopenko, Boschetti and Ryan, 2007; Grisogono, 2008), as shown in Fig 3–14.

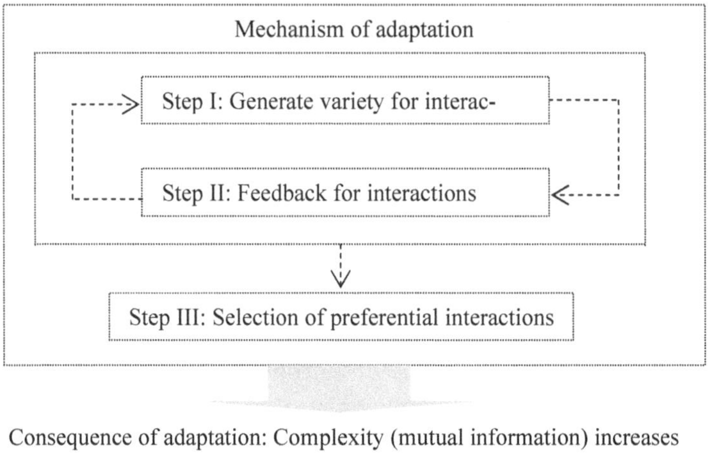

Figure 3–14 The mechanism of adaptation constituted by three essential functions

In complexity science, the term adaption is defined as a desirable property of open systems in the context of the host environments (Ho and Saunders, 1979; Grisogono, 2008b). During the evolution of an open system, for example, adaptation can be classified as global adaptation and local adaptation (Epstein, 2006). Global adaptation is an adaptive response of the system to the stimulus from its host environment, including inputs, gradients and other contextual dynamics to initiate the evolution. Local adaptation is adaptive activities of the components within the system at the macroscopic level. Suppose a system is constituted by elemental components, for example, agents or actors, those agents will internally interact with each other in response to global adaptation (Epstein, 2006), which respectively responds to microscopic self-organisation in structure and macroscopic performance of the complex system. In this sense, local adaptation can be considered as dissolution of global adaptation by a population of agents within the system. Over the time of evolution, a global coordinated performance at the macroscopic level will be generated as global adaptation to fit the specification imposed by the context of host environment, which is facilitated through the process of local adaptations within the system. In other words, the macroscopic phenomena of an open system is determined by its self-organising structure of microscopic configuration, the dynamics of which is possibly computed (Epstein, 2006).

3.3.4 Summary

In summary, the model of open systems evolution can be specified in major mechanisms: one is the macroscopic mechanism of evolution through transitional phases from equilibrium, close-to-equilibrium and far-from-equilibrium, towards the non-equilibrium steady state, which is indicated by the value of entropy; another is the microscopic mechanism of self-organisation by self-organising internal structure for an optimal distribution of energy and resource to minimise entropy till the emergence of order, or organised complexity. In brief, the macroscopic evolution is corresponding to the microscopic self-organisation over the open systems evolution, for the optimisation of open system's interrelation to the host environment. A brief modelling of open system evolutions in a spatiotemporal field is generalised in Fig 3–15.

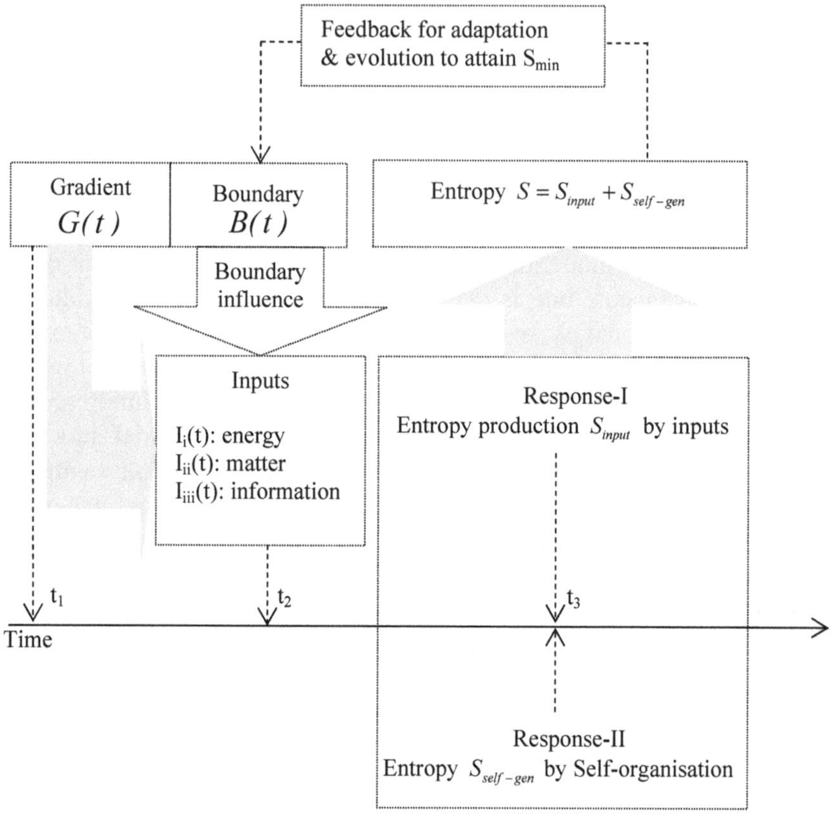

Figure 3–15 Modelling open systems evolution in a spatiotemporal field

3.4 Complexity of open systems evolution

As generalised in gas theory by Boltzmann (1964), all the possible internal agents within the system do not have an equal share in the total resource and energy. Thus, an optimal distribution of resources and energy within the system, which arrives at the completion of open systems evolution, is non-equilibrium distribution with unequal shares of resources by the internal structure. It is unnecessarily an equal share of resource. In other words, it is stability at non-equilibrium state of an open system which forms the completion of open systems evolution; equality is not the threshold to determine and to evaluate the evolution of an open system; equality does not infer stability, and stability does not come from equality.

One of options for an optimal distribution is maximum entropy probability principle which results in the broadest distribution of input resources within the open system, a diverse outcome of open systems evolution for the stable compatibility of the open system with the host environment (Shiva, 1993: 73; Hagan, 2001). This complex and optimal outcome of open systems evolution is defined as complexity of the open system after evolution. The completion of open systems evolution with a non-equilibrium distribution of resource in complexity and diversity also implies an ethical value in the view of stable social systems in Eastern religion and philosophy, which states maturity in tranquillity as the desirable and optimal outcome of evolution (Li, 1986).

3.4.1 Growth of complexity in open systems evolution

The non-equilibrium thermodynamics of open systems is argued the mother of all sciences of complexity (Schneider and Sagan, 2005), specified as follows:

- the complexity of non-linear systems (Nicolis and Prigogine, 1989) refers to any small changes in initial conditions which may lead to large amplifications of the effects of the changes;
- the properties of matter at far-from-equilibrium discloses the central prevalence of the universe is instability, and
- the non-equilibrium of irreversible thermodynamic evolution for the emergence of organised complexity as order (Kauffman, 1993).

In brief, it is the possibility of loss of stability and prediction under certain conditions that opens the opportunity of transitions leading to internal differentiation and complex behaviour. For example, in the case of only one input to stimulate the open systems evolution, via developing a complex structure and bringing up an order emerging in the system, as shown in Fig 3–16, by changing each step of this chain, complexity may be produced.

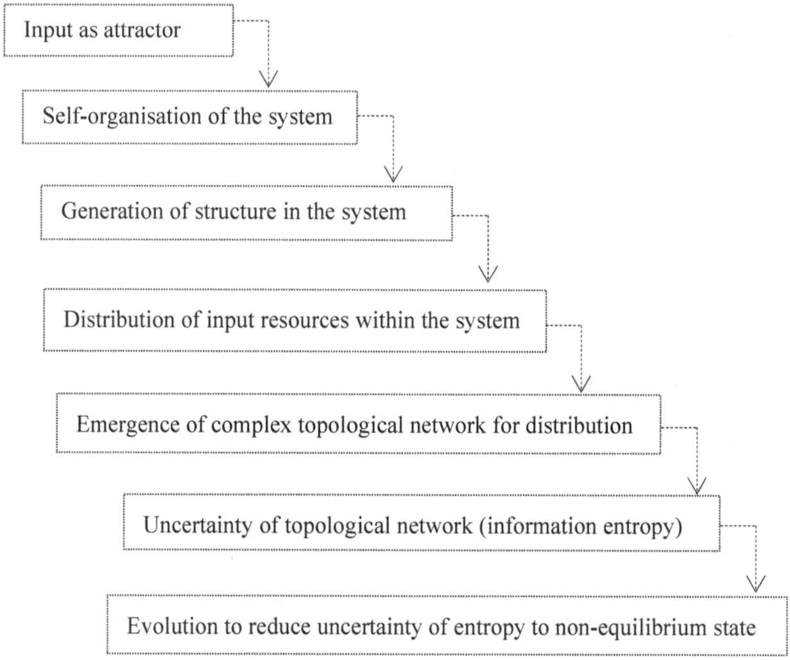

Figure 3–16 The track of complexity causality in open systems evolution

In addition to one input stimulus, if some other gradients exist between the open system and the host environment, new inputs will become attractors to initiate various evolutions for various "orders" in response emerging in the system. In consequence, new kinds of complexity will be generated, as multiple evolutions caused by multiple inputs, which is normally named as co-evolution in biological organic systems (Kauffman, 1993; Adami, 2002). In co-evolution, each order is generated in specific mechanisms responding to each input, as shown in Fig 3–17, which result in a highly sophisticated structure, and a high level complexity of the system, reinforcing the autonomy of the open system for successful survival within the host environment.

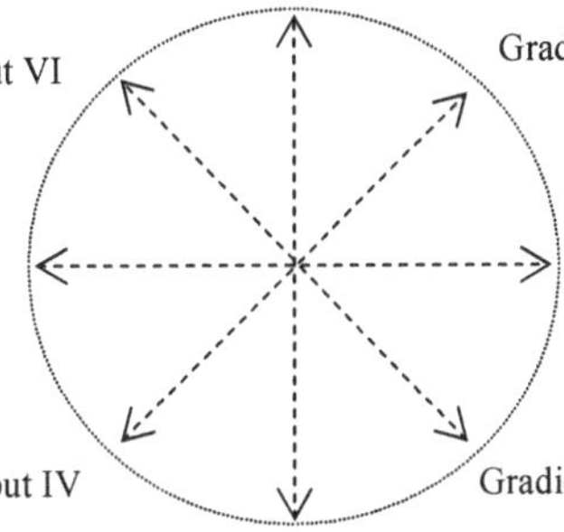

Figure 3–17 A spectrum of gradients and inputs for co-evolution of open systems

In summary, the most idealistic open system has largest capacity for taking in all kinds of inputs in large amount, developing the most highly ordered structure, resulting in the least minimal value of entropy. Such an open system certainly is the most desirable and sustainable in terms of its negative impact upon the host environment. During such a dynamics, the boundary is a crucial condition which affects the entropy dynamics, for the development of the structure and the final emergence of order. Such complex co-evolutions of open systems can be found in biological systems. For instance, the nervous system of human kind, which is utilised for efficient distribution of the pressure, energy, matter and information flows through the human body. Another example is trees, which have complex branches for effectively absorbing and distributing the inputs of energy and matter from the natural surrounding.

3.4.2 Complex performances of an open system after evolution

After the emergence of order, the open system is able to have system level interactions with other systems in the host environment if they exist (Prokopenko, Boschetti and Ryan, 2007). New iterations and cycle can repeat by following the rules of adaptation and evolution at a larger scale. At the macroscopic level, the performance of the system will possibly differ from the performance before the evolution, for the reason of the individual agent's evolution as a microscopic distribution of the macroscopic evolution of the system. At this scale, the performance of a system is argued as a coupling performance (Epstein, 2006) of the structure of the system and the state of host environment, represented as Equation (3–21).

$$\begin{aligned} \textit{The performance of an open system} &= \textit{the structure of the system} \\ &+ \textit{the state of host environmnet} \end{aligned} \qquad (3\text{–}20)$$

It is generalised that any change either in the host environment or in the structure of the system will affect macroscopic performance of the system at the system level interactions. For example, by changing the distribution and allocation of input resources sustained with a structure of the system, it is possible to adapt and optimise the on-going performance of the open system, especially to accelerate the minimisation of entropy. In the example of human civilisation, the liberty and individualisation in the modernisation of human civilisation, culture and ideology is progressing by reforming in a much broader distribution of resources to the individual constitutional units within the system of human society through the global structure, thus human civilisation evolves towards a higher level organisation compared with prior low level civilisations.

3.4.3 A general approach to the complexity of open systems evolution

Therefore, a general approach to identify the complexity of an open system as the consequence of open systems evolution from initial high entropy state to final low entropy state for the emergence of order is summarised as follows:

- Identifying a concrete meaning of order, entropy or disorder of a system. For example, order of a system can be defined as intelligence, efficiency, beauty, democracy, autonomy, satisfactory and so on; in contrast, entropy can be defined as un-intelligence, dysfunction, ugliness, authoritarian, dependence, frustration and so on;

- To attain such an order of a system, a mechanism of open systems evolution can be designed for the system, minimising macroscopic entropy production by microscopic self-organisation;
- During such an evolution, the open system will absorb the relative inputs spontaneously in the direction of the emergence of relative order.
- A complex structure will be generated within the system to facilitate an efficient distribution of the inputs resource within the system and to minimise entropy production;
- Once minimal entropy is attained at a steady state, the open systems evolution completes;
- After one gradient is fully used, the open system will spontaneously respond to other gradients for new cycles of evolution to develop new complexities.
- During such complex dynamics, the boundary condition will affect the efficiency of the entropy dynamics between the open system and its host environment.
- Through the evolution, entropy is an effective index, which indicates on-going dynamics of the system, i.e., minimised entropy implies the completion of evolution and the achievement of certain complexity of the system. The minimal entropy value is defined in terms of the compatibility of the open system and the host environment. If the value of entropy production of a system is larger than this acceptable level, then it indicates the further growth of the structure for redistributing resources within the system; or new gradients and inputs are expected for a complete new evolution of the open system.

For this open systems evolution, the existences of differential gradients, which is also called the random difference or gradients among the thermodynamic components within the isolated systems, and between the open systems and the host environment, are the essential condition for thermodynamics (Jaynes, 2003). At the completion of open systems evolution, the systems settle down with the diversity of energy topology network in maximum entropy at the state of non-equilibrium. From this viewpoint, diversity is the essential precondition and the irreversible consequence of open systems evolution, which refers to the gradients at the beginning of evolution and the maximum distribution at the end of evolution. This principle has been applied into economic theory, which states, the turbulence of individual variations is the engine that drives macroeconomic change in the direction of the equilibrium.

3.5 Comparison of open systems evolution and Darwinism evolution

3.5.1 Darwinism evolution theory

The positive implication of Darwin's evolution theory in the 19th century liberated the mankind from the belief of special creation (Prigogine and Stengers, 1984; Von Baeyer, 1998). It begins with an assumption of the spontaneous fluctuations of species, e.g. reproduction, variation, inheritance, competition and genetic programs, which are subject to natural selection for irreversible biological evolution. In general, the principles of Darwinism evolution consist of the natural selection and the fitness for the survival to maintain the equilibrium of the ecosystem (Bertalanffy, 1968; Casti and Karlqvist, 1986; Kauffman, 1993; Jaynes, 2003; Schneider and Sagan, 2005). Evolution is generalised as the process of change in forms of biological entities' life over generations during certain time course under the pressure of the environment. Then, natural selection proceeds in selective advantage via promotion or repression over the mutations of biological entities during the process of evolution. The result is the emergence of biological diversity in the ecosystem which makes the ecosystem alive and of potential for growth (Darwin, 1929; Nicolis and Prigogine, 1989). In brief, in Darwinism evolution theory, evolution of biological systems is emergent order honoured and honed by the natural selection.

However, Darwinism has been criticised as an incomplete thinking (Prigogine and Stengers, 1984; Casti and Karlqvist, 1986; McHarg, 1992; Kauffman, 1993; Ray-Jones, 2000; Schneider and Sagan, 2005). It focused on the individual organism, emphasizing the natural selection as the single important force shaping evolution; while the genes of the fittest organisms, fitness in evolutionary jargon is a mathematical measure of the survival of organisms and their ability to successful reproduce. However, the interactions between the biological organisms and the environment was overlooked, which, in fact, affects the effectiveness of nature selection and the survival of biologic systems within the natural ecosystem for their fitness. In the 20th century, a departure from the Darwinian view of natural selection of the fittest survival is Neo-Darwinism (Ho and Saunders, 1979), which suggests the creative process in evolution is the interaction of the environment and the physiological and developmental potential for adaptation inherent in organisms.

3.5.2 Open systems evolution in the Second Law of Thermodynamics

Dissipative structure theory of open systems evolution was developed by Prigogine, for which he was issued Nobel Laureate on chemistry in 1977, whose work is seen by many as a bridge between natural science and social science. However, the whole picture of evolutionary thermodynamics in terms of the Second Law of Thermodynamics was not fully configured without the interpretation of open systems' self-organisation microscopic for the emergence of order macroscopically, in complex systems theory. Certainly, the theory of open systems evolution requires further research, application and justification to be explored.

In the model of living organism's open systems evolution, it is generalised (Prigogine and Stengers, 1984; Schneider and Sagan, 2005): organic life is an open cycling system, open to the host environment, concentrating operations of energy, matter and information communications with the host environment, and optimising these communications by self-organisation to build up order or organised complexity, meanwhile producing entropy as disorder to the host environment. In total, the open systems evolution of living organic systems results in the generation of autonomy of living organic systems for their successful survival in the host environment, as shown in Fig 3–18. It means a successful survival of living organisms is able to continuously degrade existing gradients and available inputs resources to attain the autonomy; the more autonomy they can achieve, the more they can survive the vagaries of the host environment.

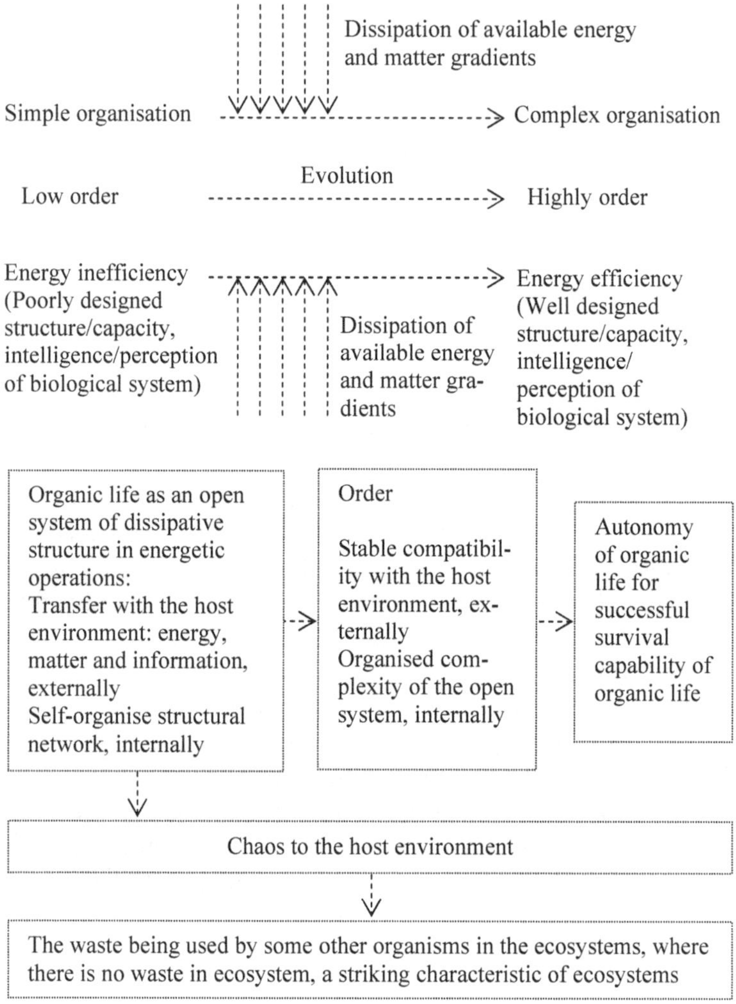

Figure 3–18 Evolution of organic life for autonomy in a model of open systems evolution

In abstract, the host environment provides the initial driving force of gradients or such as inputs of matter, energy and information, to stimulate the evolution of an open system which is a dynamic pattern of opportunities for open systems evolution (Epstein, 2006). In response, the open system evolves to a highly organised state, including adaptation and self-organisation. Finally, an optimal correlation between the open systems and the host environment is established. In brief, the open systems evolution is a combination of both self-organisation and the availability of inputs in the niche of the host environment (Lotka, 1945, 194; Schneider and Sagan, 2005), as formulated in Equation (3–22).

$$\textit{An evolutionary system} = \textit{a living organic system} + \textit{the ecosystem environment} \qquad (3\text{-}21)$$

where

- Living biological systems as open systems are spontaneous for energy capture and self-organising the energy distribution;
- The surrounding of the host environment provides contextual resources of inputs and constraints, and absorbs the waste as entropy from biological systems;
- Evolution is an entropy balance process, subject to the system's capability and the context of surroundings.

3.5.3 Comparison

The advantage of open systems evolution, compared to Darwinism evolution theory, lies in the quantitative formulation of external interactions of open systems and the host environment and internal self-organisation of open systems, both of which are in parallel to generate the relatively compatible and stable interrelations of the open system and the host environment, while these interactions of open system and the host environment were overlooked in Darwinism evolution theory. A comparison of open system evolution and Darwinism's evolution is summarized in Fig 3–19 and Table 3–5.

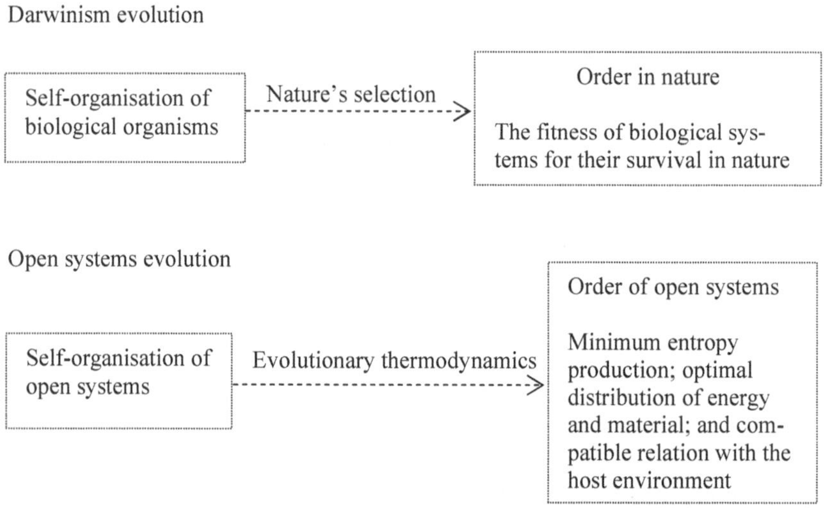

Figure 3–19 A comparison of Darwinism evolution and open systems evolution

Table 3–5 Comparisons of Darwinism evolution and open systems evolution

Items	Darwinism evolution theory	Open system evolution
Host environment	Natural ecosystems provide the opportunity for evolution of biological systems	The host environment provides the opportunity and constraints for open systems evolution: gradients, energy, matter, and information.
Internal mechanism of evolution	Generative and evolutionary of biological systems, by replicating, reproducing, mutations, and competitions	Far-from-equilibrium non-linear self-organisation of open systems
External mechanism of evolution	Natural selection: selecting via promotion or repression	Irreversible thermodynamics between open systems and the host environment for minim entropy production
Evolution consequence	The fitness of biological systems selected by nature	Order emerges at non-equilibrium thermodynamics
	Diversity of biological species	Organised complexity of the open system

3.6 Implication: a Post-Modernism world-view and a Post-Industrial pattern

The mechanism of self-organisation at a microscopic level and the mechanism of entropy decrease at a macroscopic level jointly compose the mechanism of open systems evolution, i.e. optimising a system's macroscopic performance by microscopically self-organising an adaptive structure. After the evolution, the system's performance is determined by the state of the structure and the state of the context, which jointly affect controlling and reducing entropy.

3.6.1 A Post-Modernism world-view: self-organization and evolution for the emergence of order

The fundamental principles of open systems evolution in the Second Law of Thermodynamics are composed by degrading existing gradients, interacting with the host environment through microscopically self-organising structure for optimal distribution of energy, matter and information in feedback loops, to increase the complexity of open systems, and to produce entropy in the host environment, towards the emergence of order, i.e. a macroscopic compatible interrelation of the open system and the host environment. The mechanism of open systems evolution in feedback of self-organisation and the indicator of entropy suggest a method of "evolving by learning", i.e. an open system learns from the previous experience to evolve and attain a desirable outcome, thus the system evolves smoothly from the past to the future for its sustainability. In general, the model of open systems evolution implies a new world-view of the Post-Modernism, which states the universe evolves and self-organises as an open system for the generation of its creativity and complexity.

– Theoretical implications of open systems evolution as the scientific framework revealing the new world-view of Post-Modernism are generalised as follows:

- The worldview should be evolutionary and adaptive, exhibited as a dynamic process rather than a static threshold;
- To expect any new emergence in necessity, as the essence of evolution as the progress of human civilisation;
- The system's performance is determined by the structure of the system and the state of the surroundings, the host environment;
- By generating an adaptive structure in response to the host environment, a compatible interrelation between the open system and the host environment can be achieved;
- Maximising the distribution of all available resources in the system, committed to all components within the system for minimal entropy, thus the diversity will be sustained for a long term development;
- The performance of each unit in the system is determined with reference to their distributed resources, which will affect global performance of the system with reference to the host environment.

3.6.2 A Post-Industrial pattern for environmental sustainability

The study of thermodynamics is initiated in the Industrial Age, which accelerated the thermodynamic conversion of natural resource into mechanical work demanded by the industrial progress. However, the technologies of energy use and the choices of energy use in the early industrial pattern (Huber and Mills, 2006) have induced the environmental crisis which is remarkable since the late 20th century. As argued by physicist Frederic Keffer (Rogers, 1960), "the early industrial revolution involved energy, but the automatic factory of the future is an entropy revolution". It suggests, in terms of the Second Law of Thermodynamics, the initial industrial pattern follows the pattern of closed systems or isolated systems, without an effective mechanism to optimise the energy and materials resources use in the model of open systems evolution, to minimise the negative impact of entropy upon the host environment of nature. To resolve the environmental problems caused by the unsustainable industrial pattern, the model of open system evolution therefore provides as an alternative Post-Industrial pattern for environmental sustainability.

In brief, the model of open system evolution in the Second Law of Thermodynamics provides a theoretical framework of sustainable design and development. The order of open systems evolution, i.e. the compatible coexistence of an open system and its host environment after evolution, can be interpreted as a sustainable symbiosis of buildings and cities with their host environment of the natural ecosystems. This order in buildings and cities can be realised by thermodynamically transferring energy, matter and information with the host environment of natural ecosystems which is supported by microscopically self-organising an internal structure for optimal distributions of energy and resource use over the life cycles of provision, production, transport and consumption, for macroscopically sustainable performances of buildings and cities in the context of natural ecosystems.

Part III
The Application of Open Systems Evolution to Sustainability

Chapter 4
An intelligent model for sustainable design in open systems evolution

The specifications of the Second Law of Thermodynamic and the universal generality of open systems evolution in both natural and human worlds suggest the thermodynamic relationships of the built environment and the natural ecosystem can also be interpreted in terms of thermodynamics laws (Marchettini, Pulselli, et al., 2006). In general, the built environment is considered as a form of distillation, a highly ordered material environment constructed with a highly ordered power, through the processes of refining, processing and purifying energy and material resources (Huber and Mills, 2006).

However, conventionally designed in the linear industrial pattern in modern age, buildings and cities are isolated or closed systems in terms of their thermodynamic interrelations with the natural ecosystems, without any efficient mechanism or facility automatically and spontaneously optimizing in feedback of their environmental performances. For example, in the mechanics of energy and resource use in constructional and operational services, buildings and cities take use of available energy and material resources from the natural environment to meet the microscopic environmental demands of end-users, resulting in the negative environmental impacts of waste inefficiency and pollution emission, and contributing to global warming and climate change. These negative impacts are generally called entropy in metaphor with the basis of guilt and negative sense (Ingersoll, 1996; Moore, 2006).

Thus, the model of open systems evolution, which introduces how order emerges in an open system embedded in its host environment, has a significant implication for sustainable design in buildings and cities. It provides a scientific framework of buildings and cities as open systems in the host environment of the natural ecosystem, using the mechanism of open systems evolution adapting and optimizing their environmental performances for a sustainable symbiosis with nature, as order of buildings and cities.

4.1 A scientific framework for sustainable design

The new-world view of the Post-Modernism states the creativity of the universe embodies as order emerging from its evolution and self-organisation, and the nature of the universe is revealed more instable and dynamic instead of linear and predictable in reductionism of the Modernism. This new world-view also implies a new strategy for architecture and urban design (Gleiniger and Vrachliotis, 2008). Accordingly, to generalise a scientific framework of architecture and urban design, especially of sustainable design, some critical questions are specified as follows:

Conceptual configuration of sustainable design in the context:

- What is the order of design? – Design for a specific order or an optimal performance of buildings and cities in a certain context. For example, order can be defined as an optimized desirable property to meet particular criteria of form, spatial organisation and environmental evaluation; in environmental design, order means an optimized environmental performance of buildings and cities in the context of natural ecosystem;
- Which phase is the system of buildings or cities on? – A chaotic stage of dynamic evolution in far-from-equilibrium, or the order and stable stage of non-equilibrium;
- What is the status quo of buildings and cities? – Supposed the performance of the open system at macroscopic level is determined by microscopic self-organized structure internally in the system and the state of the context of the natural ecosystem externally.

Quantitative evaluation of sustainable design embedded in the context:

- How to evaluate quantitatively the performances of buildings and cities? – Indicated by some referential index in each particular context, such as entropy of buildings and cities indicates the negative impact of their environmental performances embedded in the context of natural ecosystems, in other words, entropy symbolises the dysfunction of environmental performance of buildings and cities;
- Does the system need further evolution indicated by entropy? Is the system fully evolved in the mechanisms of open systems evolution, under-evolved, or un-evolved in the direction of reducing entropy of the system;
- Optimization of environmental performances of buildings and cities:
- Which system and structure are more adaptive and viable for desirable performances of buildings and cities, e.g. ecological sustainability in the context of natural ecosystem?
- How to evolve an optimization of system performance? – In the mechanisms of open systems evolution by macroscopic evolution and microscopic self-organisation.

4.2 Intelligent design in the model open systems evolution

Intelligent design was anticipated by Darwinism (Jaynes, 2003). The exquisite, complex structure in nature could have evolved by accumulated random mutations and by natural selection in terms of Darwinism evolution theory.

4.2.1 Intelligence of open systems evolution

The mechanism of open systems evolution is specified as follows: an open system is able to spontaneously take use of available energy and material resources from the host environment due to the existing gradients, i.e. the differentials of chemical, physical, temperature, density and pressure between the open system and the host environment; by taking use of these inputs, the open system evolves from the initial chaotic far-from-equilibrium state to a highly ordered non-equilibrium state when the system is stable and compatible with the host environment; the evolution of an open system in the host environment is indicated by the change of entropy produced by the open system, which changes from the initial maximum value to the final minimal value at the non-equilibrium state of the open system.

Thus, the intelligence of an open system is generalised as such: an open system able to adapt to the host environment in spontaneity for its evolution; during the evolution, the open system is aware of its ongoing performance and its impact upon the host environment which is indicated by entropy, able to self-organise a highly ordered structure for resource distributions, in order to optimise its performance with minimal negative impact; in result, a compatible interrelation of the open system with its host environment or other counterparts is generated for sustainable co-existence in a long term.

Applying the intelligence of open systems into buildings and cities, an intelligent model of buildings and cities as open systems refers to be intellectually sensitive, adaptive and responsive to the environmental context (Leach, 2009). By adapting to the environmental context and evolving in the mechanism of open systems evolution from the chaotic stage towards an order state, open systems of buildings and cities would be able to optimize their environmental performances, with minimal dependences on energy and material resources, with minimal, neutral and even positive environmental impacts on the natural ecosystems, which is technically embedded and constrained by the natural context.

4.2.2 Sustainable design in the intelligence of open systems evolution

In the model of open systems evolution, the concept of sustainability is generalised to attain a balanced ecological interrelation between the open system and the host environment through an evolutionary process. Applied into buildings and cities, sustainable design as an alternative design standard in the coming decade is proposed not only to design for energy and resource efficient and pollution reduction, responding to the social-economic-environmental imperatives, but more significantly, sustainable design refers to design with an intelligence, being aware of its environmental impact, sensitive and adaptive, automatically optimising its environmental performance and impacts upon the natural environment. In summary, sustainable design defined in the model of open systems evolution is a highly intelligent design.

According to open systems evolution's mechanisms of self-organisation, adaptation and optimisation, environmental performances of buildings and cities are proposed to be adapted and optimized by actively ecological interactions with the natural environment, in the form of thermodynamic exchanges of energy and material resources with the natural ecosystem. Thus, buildings and cities can work as ecologically productive systems as part of the natural ecosystem for the potential sustainable symbiosis between them. In summary, the nature of sustainable design is identified as an optimisation of environmental performances of buildings and cities in the mechanism of open systems evolution, for environmental sustainable interrelations of buildings and cities in the context of nature.

4.3 Conceptual configuration of intelligent sustainable design

In the model of open systems evolution, sustainable design in buildings and cities can be configured in scheme by defining key terms of order, the host environment, open systems, entropy, etc, as shown in Fig 4–1.

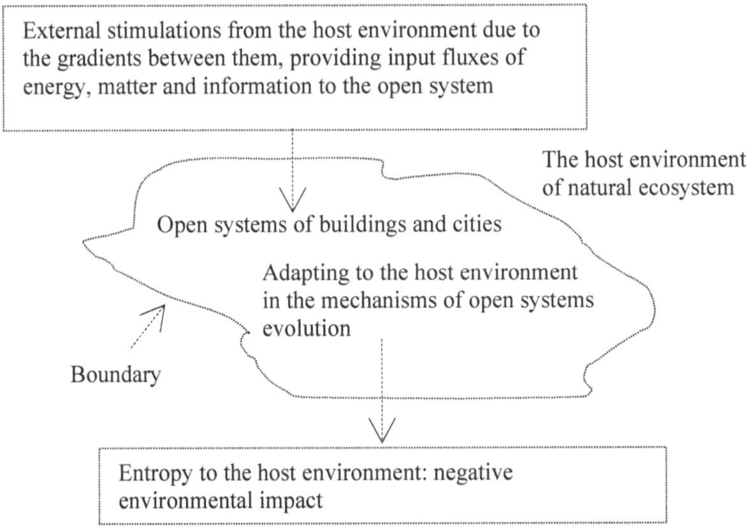

Figure 4–1 A primary scheme of sustainable design in the model of open system

4.3.1 Order of buildings and cities in the environment sense

Historically, there were several transformative and controversial identifications of order in the built environment, each of which is responsive and also constrained to the time ideology and perception. For example, traditional architects have felt impelled to create new social orders through new formal orders, geometric orders in architecture and urban design. An ordered architecture has represented an ordered and meaningful society set in an ordered and meaningful universe, with the relationship between signifier and signified fixed, closed and stable; in the modern urbanisation, to control the unplanned growth, for example, with their ideal cities, Renaissance architects intended to overcome the chaotic medieval town; in the industrialisation of the 19th century, networks of streets and streetcars formed the nervous system of spreading national states (Mainzer, 2008); Haussman intended to render Paris fit for the 19th century, while Le Corbusier wanted to rationalize it for the 20th century (Hagan, 2001).

In terms of the model of open systems evolution, order or organised complexity of a city can be identified by classifying a series of categories as the microscopic structure of a city. For example, if the accessible mobility of a city is defined as a desirable "order", the "inputs" can be identified as the diverse urban activities; thus, entropy of urban mobility can be formulated and computed in the mechanism of open systems evolution for the optimisation of urban mobility. From the environmental perspective, by efficiently self-organising energy and material fluxes over their life-cycles, entropy, which means the negative environmental impacts upon the context of the natural environment, can be minimised, and even neutral or positive contributions to the natural eco-system can be produced. Thus, order of buildings and cities in the environmental design is defined as ecologically compatible interrelations of buildings and cities with the natural ecosystem.

4.3.2 Nature as the host environment for buildings and cities

Considering the natural environment as the context of buildings is one of basic principles in vernacular architecture in the Pre-Industrial age. In the Post-Modernism, contextualism has also been emphasised as a new design strategy to re-introduce the overlooked natural and other cultural concerns into the simplified functionalism of Modernism (Venturi, 1950; Gleiniger, 2008). In the model of open systems evolution, the natural ecosystem is an essential dimension in sustainable design, works as the host environment of buildings and cities, in which buildings and cities will optimise their environmental performances in the mechanism of open systems evolution, establishing ecological compatibility between them and reinforcing the environmental complexity of buildings and cities.

4.3.3 Open thermodynamic relations of buildings and cities with nature

A building and a city normally take use of primary energy and material resources from nature, which are converted into useable and organised energy for micro-climate control and other operational services, and emit pollution emission to the natural environment. The complex thermodynamics of urban territory and buildings have been investigated and configured by Pulselli et al (2005) and Marchettini et al (2006), as shown in Fig. 4–2 and Fig. 4–3. Thus, buildings and cities can be identified as open systems of dissipative structure, which continuously exchanges flows of energy and material with the ecological surroundings as supportive services, to survive, develop and increase in complexity.

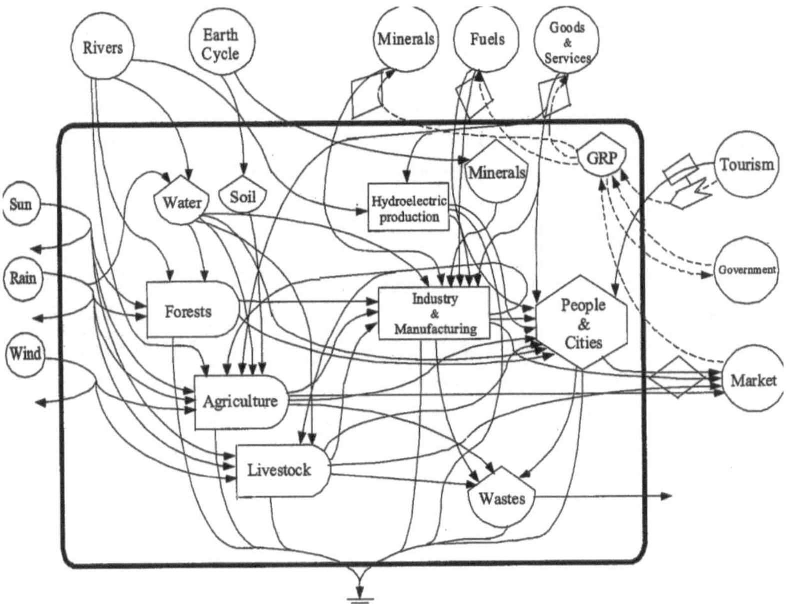

Figure 4–2 Energy and resource flows and transformation in an urban territory[25]

25 Diagram source: Marchettini, N., Pulselli, F. M. and Tiezzi, E. (2006) Entropy and the City, in Mander, Ü., Brebbia, C. A. and Tiezzi, E. (eds) The Sustainable

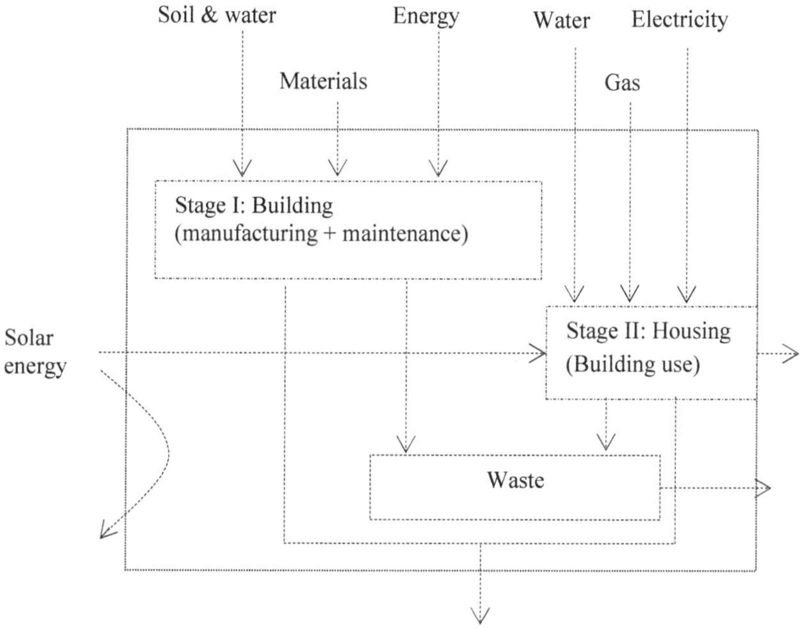

Figure 4–3 Energy and resource flows and transformation of a building system

In the model of open systems evolution, an open thermodynamics network of buildings and cities includes both external thermodynamic interrelation with nature and the internal thermodynamic distribution network within buildings and cities. By continuously taking diverse categories of energy and resources, not only the existing prevalence of unrenewable energy and resources, e.g. fossil fuel and electricity, but also the renewable ones, e.g. solar, wind, geothermal and etc, it is possible to reduce the negative environmental impact of greenhouse gas emission caused by fossil fuel use in the building industry.

City IV: Urban Regeneration and Sustainability, Tallin, Estonia, Southampton, UK, Boston: WIT. Copyright permission of diagram is granted to the reproduction in this book by WIT Press.

4.3.4 Mechanism of open systems evolution for optimisation

In addition to the configuration of open thermodynamic relationship with nature, the mechanism of open systems evolution, including both macroscopic evolution and microscopic self-organisation, will optimise the environmental performances of buildings and cities. The metabolism of thermodynamically open relationship of architecture with the natural environment is initially proposed in a model of evolutionary architecture (Frazer, 1995), in which, architecture is argued to maintain its stability with the environment by negative feedback interactions and promote evolution in its employment of positive feedback. The non-linear mechanism of open systems evolution replaces the linear industrial design pattern of energy use in buildings and cities, in which nature is considered as source to buildings and cities providing energy and material resources, while nature is also used as sinks for the emission from buildings and cities.

4.3.5 Sustainable topology of energy and resource uses

In order to optimize the environmental performances towards desirable order of buildings and cities, such as minimised negative environmental impacts, neutral or even positive impacts, it is necessary to follow the mechanism of open systems evolution and to self-organise an internal structure of energy and material use in buildings and cities. An alternative optimal distribution of resource within a system is maximum entropy principle, in which, open systems evolution can be completed by self-organising frequency distribution of individual variations in the system (Jaynes, 2003) with minimal entropy production.

Applied in sustainable design, an optimised topological network or structure of energy and material resources can be generated in buildings and cities in maximum entropy principle. It requires the probability distribution of energy and material usage in buildings and cities which are available in the context of natural environment. By computing and minimising entropy of negative environmental impact, a topology of energy and material use responding to sustainable environmental performances of buildings and cities can be approached. Several existing strategies can be alternative for the sustainable topology of energy and resources use, such as resource autonomy and recycling for energy self-sufficient, self-contained and self-organising over their life-cycles. New strategies will be generalised in the following.

4.3.6 Entropy of buildings and cities in the environmental sense

The basic meaning of entropy in thermodynamics science refers to unavailable and wasted energy during the thermodynamic transformation, a dysfunction of the system; in the open systems evolution, entropy indicates on-going evolution of open systems, from maximum value to the minimal value at the state of non-equilibrium. Applied to buildings and cities, primary questions regarding to a comprehensive identification of entropy are raised as follows:

- A qualitative identification of entropy in the built environment: what is entropy in buildings and cities? – A thermodynamic consequence of buildings and cities with the natural environment through energy and material resource uses.
- A quantitative approach to measure entropy in the built environment: how to measure entropy in building?
- The implications of entropy in buildings and cities: why is it important to study entropy? – A mechanism of open systems evolution indicated by entropy, for adaptation and optimization of environmental performance and impact, through technical innovations in material, structure, and even conceptual design.

4.3.6.1 Entropy as a conceptual metaphor of negative environmental impacts

The concept of entropy in the built environment is a metaphor of negative environment impacts (Ingersoll, 1996; Knowles, 2006), i.e. the residual and waste during the allocation and distribution of energy and resources, emission pollution and other negative impacts produced by energy flows and resources used in buildings and cities to sustain their metabolism over their life-cycles. The term of "entropy" was introduced into ecological architecture in architectural theory in the 1860s when "ecology" were both coined (Ingersoll, 1996); by the 1920's, the doctrines of energy economics was promoted in architecture as ethical and economic imperatives (Moore, 2006) to expend energy as efficiently as possible, in technological innovations along with the development in natural science. In 1960's, Olgyays (1963) applied the concept of entropy to architecture. However, the major appreciation of the salience of Claussius' theory in buildings and cities design started in 1970s when energy crisis emerged globally (Moore, 2006).

One of the most typical alternatives of the high-entropy path in urban organisations is argued as the Garden City (Ingersoll, 1996; Hagan, 2001), followed by sprawling Broad-acres City and high-rise based Radiant City (Steele, 2005).

4.3.6.2 Entropy as a technical indicator of negative environmental impact

According to its original meaning in thermodynamics and the metaphoric meaning in the built environment, in environmental management, entropy works as a synthetic indicator of the negative environmental impact of buildings and cities for efficient evaluation of their environmental performances (Moore, 2006; Pulselli and Pulselli, et al., 2006). In this optimisation, entropy can work as a quantitative index for adaptation, evolution and optimisation of environmental performances, as shown in Fig 4–4.

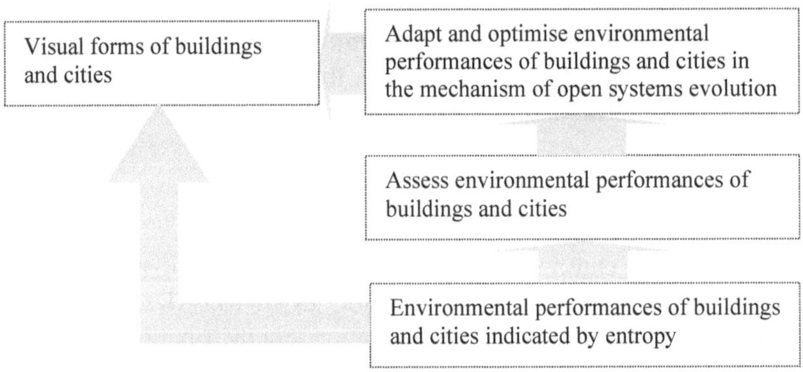

Figure 4–4 A proposition of optimising environmental performance indicated by entropy

An alternative approach for evaluating environmental performance of a building is how long a building could reach the state of minimised negative, neutral or positive impact within its life-cycle, which includes its production, construction, operation and demolition phases. If the life-cycle of a building is ***n*** years, a building reaches the state of minimised negative environmental impact in one year, its environmental performance grade is *(n – 1)*; if a building reaches the state of minimised negative environmental impact in 10 years, its environmental performance grade is *(n – 10)*; if a building never reaches the minimised negative environmental impact state within its life-cycle, its grade is zero, which means the building is negative, without any positive ecological contribution to the environment within limited timeframe.

Several methods have been devised for the quantification and the analysis of energy and resource use in building system, including environmental accounting, eco-auditing (Lang, 1995) and life-cycle analysis (Field, 1999), environmental accounting (Odum, 1996), with the physics basis of the Second Law of Thermodynamics. In particular, in emergy analysis (Odum, 1983; Pulselli, Pulselli, et al., 2006), buildings work to provide specific services and maintain their performances constant in time with respect to variable context conditions, i.e. clime, temperature, humidity, sun irradiation, atmosphere, etc; therefore, buildings management is to take care of the interactions between buildings, i.e. non-living inert structures, and their living context, i.e. the environment and human beings. This dynamic network of relations could depend on people movements and exchanges of goods and services; it has the form of materials and energy flows. One of experimental results of building materials evaluated in emergy analysis is illustrated in Table 4–1.

Table 4–1 Emergy analysis of raw materials flows in the building[26]

Raw materials	Specific emergy (sej/kg)	Ref	Emergy (sej)	%
Concrete	1.79E+12	a	5.49E+18	42.64%
Tile and Brick	3.68E+12	b	2.83E+18	21.99%
Waterproof sheathing	8.85E+12	b	2.37E+18	18.44%
Stony materials	2.44E+12	d	7.54E+17	5.79%
Steel	6.97E+12	b	5.39E+17	4.18%
Plaster	3.29E+12	e	3.79E+17	2.94%
Painting	2.55E+13	b	2.91E+17	2.26%
Copper	1.04E+14	c	9.20E+16	0.71%
PVC	9.86E+12	b	5.71E+16	0.44%
Mortar	3.31E+12	b	3.45E+16	0.27%
Alloy	2.13E+13	b	3.17E+16	0.25%
Wood	2.40E+09	d	1.17E+16	0.091%
Glass	2.16E+12	e	1.69E+14	0.0013%
Total building materials emergy			1.2873E+19	100.00%

4.3.6.3 Entropy as Nature's technique for balance in ecological design

The thermodynamic principles in the Second Law of Thermodynamics, such as heat goes from hot to cool matter downhill, degradation of gradients in the model of origin-of-life to process the two-temperature thermal cycle and drive self-replicating transformation of chaos to order in energy-capturing cycle between two systems for the spontaneous thermodynamics exchange, are briefly called "entropy as na-

26 Table source: Pulselli, F. M., Pulselli, R. M., et al. (2006) Environmental accounting of buildings: outcomes from the emergy analysis, in Mander, Ü., Brebbia, C. A. and Tiezzi, E. (eds) The Sustainable City IV: Urban Regeneration and Sustainability, Tallin, Estonia, Southampton, UK, Boston: WIT. Copyright permission of table is granted to the reproduction in this book by WIT Press.

ture's balance for technique" (Bisch, 2001), as shown in Fig 4–5. These have been applied as technical innovations in ecological design to improve the thermodynamics of buildings as organic systems, to reduce energy consumption, to spontaneously capture available energy and resources between the buildings and the external natural environment, e.g. taking use of temperature and pressure differentiates and gradients, to optimise the consumption and distribution of energy and resources in a manner with minimised negative environmental impact. For example, applied in the case of ceiling activation system (Bisch, 2001), buildings are designed to adapt the internal thermal energy use and energy saving, to reduce the dependence of conventional HVAC use. In brief, the material association is actively taken into thermodynamics of the building as a living ecosystem, in aim for ecological balance and sustainability.

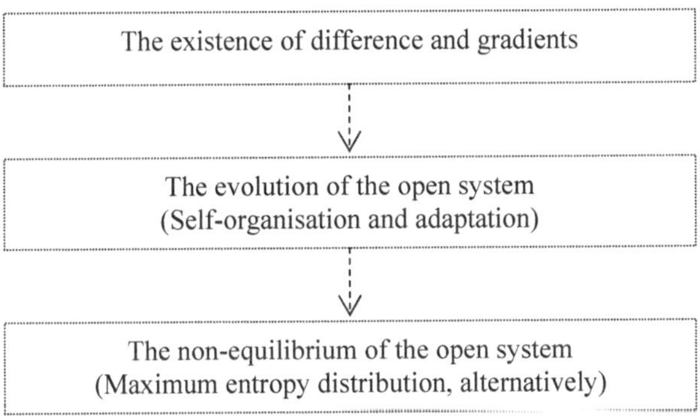

Figure 4–5 Entropy, Nature's technique for balance

4.4 Parametric design of intelligent sustainable design

4.4.1 Parametric design scheme of open systems evolution

The environmental performances of buildings and cities can be formulated in a mathematical parameters, which include the local climate context, the end-users' demands and feedback, the available energy and material resources, the distribution and consumption of energy and material resources, the environmental impact of negative energy waste and emission or positively self-generated energy, and the available technology for energy and resource control, etc. These parameters can be configured by translating and classifying into the model of open systems evolution which is composed by the open system, the boundary, the host environment, the gradients, entropy, the mechanisms optimising the topology of energy and resource, the constraints and the opportunities, as follows:

- Operators or drivers of the evolution: the end-users' brief such as micro-climate control;
- Open systems: the environmental performances of buildings or cities;
- Constrains and boundary of the system: the conflicts between the climatic context and the end-users' demands and comfort; the availability of information, technology and techniques for environmental performance;
- The host environment: the local environmental and climatic context;
- Open systems evolution: the input and output of energy and material resources across the boundary of a building or city system;
- Entropy: the irreversible environmental impact such as emission and waste, and positive contributions such as self-generated and self-contained energy and other resources.

These parameters will follow the mechanism of open systems evolution, adapting and evolving towards the positive outcome of a highly-ordered organisation of energy and material resources for sustainable environmental performance of buildings and cities, generalised as follows:

- To configure the macro-state performance of the system, and micro-state structure of the system;
- To distinguish the differentials of macro-state parameters and micro-state parameters in the chain of causality-and-effect;
- To identify the contextual condition and constraints, the natural surrounding, as the host environment for the open system, i.e., available resources, and boundary conditions, available techniques;
- To optimise the macroscopic performance of the system in the mechanism of open systems evolution, by self-organizing microscopic structure, subjected to condition and constraints in the host environment;
- To find the optimal distribution of energy and material resources for micro-state configuration in maximum-entropy probability principle, subject to the available information and resources;
- Optimization will be repeated in feedback loops, to identify an acceptable threshold of sustainable environmental performances of buildings and cities in the context constrains of natural ecosystems, for the minimal value of entropy;
- According to the minimal value of entropy, the desirable state of non-equilibrium of the open system, which refers to an optimised environmental performance of buildings and cities with the natural context will be established;
- This sustainable environmental performance of buildings and cities non-equilibrium state with its microscopic topology of energy and resource distribution in the systems of buildings and cities will constitute a set of evaluation criteria for sustainable design in a reference to the natural context;
- According to these attained criteria of sustainable design, the status quo of other systems of buildings and cities, their macroscopic performance and microscopic organization will be evaluated, predicted (Li and Vitányi, 2001) and optimised towards desirable sustainable performance in fitness to the context of natural environment.

In terms of the model of open systems evolution, the scheme of open systems evolution of buildings and cities to optimise their environmental performances can be constituted in the steps of capturing, distributing, transferring, self-organizing and optimising, generalised as follows:

- Spontaneous change and conversion in chemistry and biochemistry for the occurrence of thermodynamic change between the open systems of buildings and cities with the natural ecosystem, due to the gradients, e.g. temperature differences;
- Emergy analysis for the conversions of energy and matter into one unit as microscopic parameters, including human demands from the end-users, to configure a holistic system of environmental performances of buildings and cities in mathematical languages (Jaynes, 2003), as shown in Table 4–2;
- Optimisation by computing the microscopic parameters in the mechanism of open systems evolution to attain an alternative optimal distribution of energy and material resources in terms of maximum-entropy probability principle towards minimised negative environmental impacts.

Table 4–2 Parameter design in probability theory for optimisation design[27]

Configuration of parameters	Value of the parameters
Parameter I ··········	··········> X_1
Parameter II ··········	··········> X_2
Parameter III ··········	··········> X_3
… ··········	··········> …
Parameter N ··········	··········> X_n

27 (Jaynes, 2003)

4.4.2 Information control system for adaptation and evolution in algorithm

This parametric scheme of sustainable design constitutes a holistic information system of environmental performances of buildings and cities, as shown in Fig 4–5. Virtually, by this information system, environmental performances of buildings and cities can be precisely described, simulated, analysed, controlled, adapted and optimised in the mechanism of open systems evolution. The organisation of energy and material resources include capturing, production, circulation and distribution, consumption and emission, with the active involvement of the end-users and the sensitive response to the local environmental context within their life-cycles. The possible achievement of sustainable performance by this building information modelling is subjected to the constraints of the end-users' demand and the local environmental context.

Several mathematical tools can be employed to describe the open system model of buildings and cities. For example, probability theory can be applied to describe the distribution of resources in maximum entropy probability principle. Quantum mechanics can be applied to describe uncertainty and the limits of precise measurement such as unattainable data of embodied energy and some operational energy, and tensor calculus will be applied to describe the incessant dynamics of the evolution of open systems of buildings and cities. By assumptions, it is possible to design a controllable system with a finite number of measurable and observable states constituted by finite measurable parameters with available technical and technology.

In summary, the information modelling of parametric design for sustainable design in buildings and cities in the model of open systems evolution will provide sufficient information of ongoing performance of buildings and cities, to facilitate the mechanism of open systems evolution for the optimisation of their environmental performances. Thus, not only the efficiency of energy and resources use in the life-cycle of buildings and cities will be improved to minimise their negative impacts over life-cycles, but also possibly, the positive impact will be realised qualitatively. By these active interactions with the natural ecosystem, it is possible to realise a sustainable symbiosis between buildings and cities with the natural ecosystem. Therefore, buildings and cities will work as ecological productive facilities, with material's ecological performance and capacity, and structural associations.

4.5 Implications of open systems evolution in sustainable design

The paradigm of open systems evolution is a guiding principle in Nature that seems specifically designed for sustainability. Learning from Nature, several implications of open system evolutions in architecture are generalised as follows, which is briefly summarised as that sustainable design via an open system model is a synthesis of arts and sciences.

4.5.1 A holistic scheme for system sustainability of man, nature, buildings and cities

Based on the establishment of open thermodynamic relations of buildings and cities with nature, the manipulation of the mechanisms of open systems evolution into the life-cycle operation of energy and material resource use in buildings and cities, and with the involvement of human dimension, an integrated scheme of system sustainability, embracing nature, buildings and cities, and man, can be configured, as shown in Fig. 4–6. In this holistic scheme, the participation of end-users and the concern of the environmental impacts are interactive positively into sustainable design of building and cities. In other words, through the medium of buildings and cities as man-made interventions into nature, an ecologically positive relationship of man and nature can be achieved in the model of open systems evolution. In our current industrial age, this human interdependence on nature is significant flor the long term sustainable development. It implies, compared to the greatness of Nature, how powerless human being is after more than two hounded years' aggressive and desperate hunting, fighting, snatching in the industrial pattern.

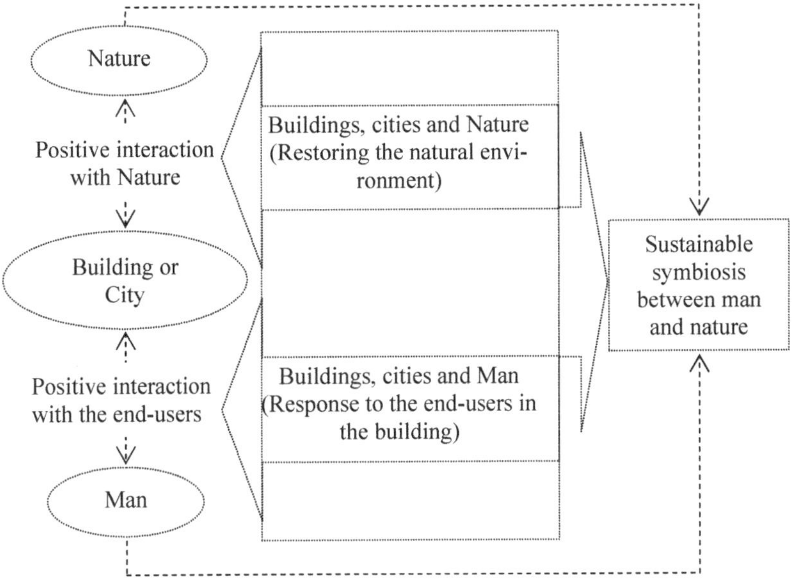

Figure 4–6 A holistic design scheme for system sustainability

4.5.2 Positive ecological design in the model of open systems evolution

In sustainable design, the positive implication of the model of open systems evolution is proposed that buildings and cities will function as open ecological systems to convert the primary energy and material resources into highly organised built environment without negative impacts upon nature. Theoretically, in the model of living organic open systems, buildings and cities are proposed to evolve from initial entropy positive state to entropy balance state of a non-equilibrium state, and possibly to a positive environmental contribution state, however, which is subjected to the constraint and conditions of the context, including technical and natural contexts.

4.5.3 Emergence of ecological order in buildings and cities

In terms of the Second Law of Thermodynamics to open systems, the new arrival of material and energy in the system moves naturally down gradients toward an equilibrium condition. The emergence of a system is defined as the system to be pushed away far-from-equilibrium when matter and energy enter (Allen, 2001). Flux down these gradients causes the system to become more elaborate to complete the emergence. New structure appears with the elaboration of emergence, so does new relationships of control and constraint, the creation of organization. In the metabolism of buildings and cities, buildings and cities digest the physical or bio-chemical inputs and produce the outputs of energy and material in feedback loops of adaptation and optimisation. In the end of open systems evolution, the negative impacts upon the natural ecosystems is minimised, and the energy and material use is highly organised in a sustainable topology, both of which constitute the emergence of order in buildings and cities in the ecological sense. It is a significant departure from the conventional metabolism of linear industrial pattern, which removes the pollutants or excess materials offsite to centralized systems, such as landfills, or sewage treatment plants.

4.5.4 A design theory of contextual architecture

The paradigm of open systems evolution can be generalised as a strategy of contextual design for sustainable symbiosis of architecture with nature. This design strategy is notionally entitled "a design of contextual architecture". Contextual architecture refers that buildings would be designed environmentally responsive to the natural context, and dynamically interactive with the natural environment for the optimisation of their environmental performances, which can be realised by information communication technologies. In the terms of information science, through "pattern recognition and learning experience" in the paradigm of open systems evolution, it is possible to realise contextual architecture, where the morphological design language, the structure, the materials, the spatial organisation, the topology of energy and resources can all be environmentally responsive to their various contexts.

4.6 Conclusion

In the model of open systems evolution, an intelligent model for sustainable design is proposed as a holistic information system, integrating buildings and cities, man and nature for a sustainable environmental symbiosis. This intelligent scheme of sustainable design is constituted by establishing open thermodynamic relations of buildings and cities with nature; adapting and self-organising their environmental performances in the mechanism of open systems evolution towards the optimal topology of energy and resource and the optimisation of environmental performances. In this intelligent model, entropy is a sensitive indicator of on-going evolution in feedback loops. The information communication technology will provide accessible channel for the buildings and cities' end-users' participation and to meet their micro-climate control of buildings and cities. In brief, the information modelling is generalised as a tripling, optimising the topology of energy and resources within the systems of buildings and cities, minimizing the negative impact of entropy to the natural environment, and satisfying the demand of end-users of buildings and cities.

An intelligent model for sustainable design in buildings and cities in a paradigm of open systems evolution is briefly summarised as follows:

- A system: an abstract method to study buildings as complex open systems, the configuration of which is embedded in the natural environmental context, imposing certain opportunities, constraints and limits for the optimisation of the environmental performance of a building.
- An indicator: buildings would be designed as open systems, which are intelligently aware of their environmental impact and adapt their environmental performance to minimise negative envi-

ronmental impact; all of the on-going process would be indicated by a quantitative index of entropy; this indicator would then suggest the possible direction of optimisation; this quantitative indicator is specifically defined regarding to the host context of nature, as stated in the model of open systems evolution.

- A mechanism: in the mechanism of open systems evolution, buildings are designed to be able to adapt and optimise their environmental performances with minimal negative impact over their life-cycles, to be environmental compatible to the host environment of the natural eco-system.
- A contextual definition: the concept of sustainable design is defined as an environmental compatibility between buildings as open systems with their natural context as the host environment, which can be achieved through complex environmental interactions between them through the mechanism of open systems evolution.
- An on-going process: the primary goals of sustainable design, i.e. the efficiency of energy and resource consumption and the minimisation of negative environmental such as pollution emission, are realised by an on-going dynamic evolution of buildings as open systems; buildings would be in an intelligence to be aware of the negative impacts, and to actively and positively adapt and optimise their performances with minimal impacts.
- Generality in application: the strategy in the paradigm of open systems evolution is applicable not only in sustainable design in architecture, but also in sustainable development in general. The interrelation of a development and its contexts, e.g. environmental context, sociological context, economic context, cultural context, geographical context and etc, can be interpreted in terms of the interrelations of an open system and its host environment; thus, in the evolutionary mechanism of open systems, sustainable development can be achieved with minimal negative impacts upon the context.

In conclusion, the model of open systems evolution is a scientific paradigm of ecological design for environmental sustainability in thermodynamics terms, a theoretical framework for dynamic control of environmental performance of buildings and cities over their life-cycles, and a technical innovation of information control system design for a holistic sustainability of man, buildings and nature. It extends and reinforces traditional ecological philosophy, by re-defining the ecological interrelations of buildings and cities with nature in the terms of open systems model and in the effective optimisation in the mechanisms of open systems evolution.

Glossary

Industrial Revolution: The rapid development of a nation's industry, especially that which occurred in Britain in the second half of the 18th century and the first century half of 19th century, in which the bulk of the working population turned from agriculture to industry. Preceded by major changes in agricultural methods which freed workers for the factories, it was made possible by the rise of modern industrial methods, with steam power replacing the use of muscle, wind, and water, the growth of factories, and the mass production of manufactured goods. The textile industry was the prime example of industrialization, and created a demand for machines and for tools for their manufacture, which stimulated further mechanization. Improved transport was needed, provided by cannels, roads, railways, and steamships; construction of these required a large labour force, and the skills acquired were exported to other countries. It made Britain the most powerful industrial country in the world but radically changed the face of Britain society, throwing up large cities (particularly in the Midlands) as the population shifted from the countryside, and causing or exacerbating a series of profound social and economic problems (Pearsall and Trumble, 2003).

Climate Change in Intergovernmental Panel on Climate Change (IPCC) usage refers to any change in climate over time, whether due to natural variability or as a result of human activity. This usage differs from that in the Framework Convention on Climate Change, where climate change refers to a change of climate that is attributed directly or indirectly to human activity that alters the composition of the global atmosphere and that is in addition to natural climate variability observed over comparable time periods (IPCC, 2007).

Ecological footprint is "the area of land and water required to produce the resources consumed and to assimilate the wastes generated by the population on a continuous basis, wherever on earth that land is located" (Rees and Wackernagel, 1996).

Sustainable Development is the development that meets the needs of the present without compromising the ability of future generations to meet their own needs (Brundtland Commission sustainable development, 1983).

Modernity: comprehensive definition of the condition of modernity captures the full scope of the kaleidoscopic character of contemporary life, which he traces to many sources. These are the great discoveries in the physical sciences, changing our images of the universe and our place in it, the industrialisation of production, which transforms scientific knowledge into technology, creates new environments and destroys old ones, speeds up the whole tempo of life, generates new forms of corporate power and class struggle, immense demographic upheavals, severing millions of people from their ancestral habitats, hurling them halfway across the world into new lives, rapid and often cataclysmic urban growth, systems of mass communication, dynamic in their development, enveloping and binding together the most diverse people and societies (Sociologist Marsall Berman, All That is Solid Melt into Air: the experience of modernity, Simon and Simon, New York, 1982, page 16).

Post-modernism/Post-Industrial Society: The term "Post-Industrial Society" introduced by social scientist Daniel Bell (1973) argues that we are in the third and final stage in the transition from an agrarian society. This transition began with the invention of steam power by James Watt in 1765 at the beginning of the Industrial Revolution, a fundamental shift from the dependence on human strength and to the power of the machine. The second stage occurred during the industrial age, when steam power was replaced by electricity. And in the third stage, "information, derived from knowledge has been encoded in comprehensive technological systems." (Bell, 2001) The developing worlds is by definition still in the process of industrialisation, going through the same social upheavals as its developed counterpart did more than two hundred years ago, and is also in the throes of the electronic revolution: which means it is experiencing its industrial and post-industrial age at the same time (Steele, 2005).

Complexity science: the definitional authority of complexity science was located indisputably in the gravitational fields of cybernetics, computer science, and information technology, quantum physics, fractals theory, chaos theory, catastrophe theory, emergence theory, nonlinear dynamics theory (Jencks, 1997). In contrast to the modern science of simplicity in reductionism, mechanism, and linear determinism, complexity science reveals the creativity of the universe in organised complexity, as the manifesto of the post-modernism (Gleiniger and Vrachliotis, 2008).

Trans-disciplinary approach for system design towards sustainability: work across the boundaries of academic disciplines, i.e. social, ecological, psychological, economic, political; communicate in many 'languages', i.e. legal, numerical, conceptual, aesthetic, with clients, collaborators and decision makers from different perspectives and backgrounds; avoid or reduce negative impacts of projects on all levels, i.e. site specific, regional and global; understand how philosophical underpinnings of design influence their decisions; Integrate different practical parameters: functional and ecological requirements, social needs, cultural values and economic constraints; work to achieve new social goals, while dislodging existing social, economic and political impediments to sustainability; think simultaneously on different scales. i.e. from the design of product component to complex urban development; dcvclop an ability to deal with uncertainty and uncharted territory, i.e. the future (Birkeland, 2002).

Integrated thinking: Energy efficiency requires to think in an integrated manner, gaining an overall view of the various energy flows involved in the act of buildings, from energy production to service industries: production energy, grey energy: losses through transport, distribution, operating energy, induced energy consumed in the course of construction (Salat, 2006).

Ecology: (1) the branch of biology dealing with the relations of organisms to one another and to their surroundings (2) human ecology: the study of interaction of people with their environment. Reference:

ecology is a relatively new science; the word is said to have been coined by E.H. Haeckle in 1869. It may be concerned with the study of a single species in its environment, or of a complete ecosystem. Aspects that are investigated by ecologists include the dynamics and distribution of populations, the interactions between species, the effects of the physical environment, and the energy flow through the system. The study of human environmental impacts is forming an increasingly important part of ecology (Pearsall and Trumble, 2003).

Ecosystem is a biological community of interacting organisms and their physical environment (Pearsall and Trumble, 2003).

Ecological design seeks to acknowledge, facilitate and preserve the interrelationship of nature and buildings, learning from nature and applying the ecology to building, a model or metaphor for sustainable building, natural system be directly incorporated to improve the functioning of the built environment, human-nature interface be managed for the benefit of both systems: optimize urban morphologies, passive solar systems and leading edge technologies in various climatic context, highest performance sustainable construction methods, systems, products and materials on the market, environmental management of buildings, eco-consumerism: energy management (Yeang, 1995)

Ecological balance means using scarce land and building resources in a way that safeguards the interests of future generations. The need to preserve the natural environment is a major sustainability issue. Land use must be treated as an ecosystem rather than as a commodity. Cities should thus be apprehended in their entirety as closed-loop ecological systems. They should be as autonomous as possible by optimizing resources reuse, with input and output levels (resources and wastes) that respect the carrying capacity of the earth (Salat, 2006).

Thermodynamics is the science of the relations between heat or other (mechanical, electrical etc.) forms of energy, and, by extension, of the relationships and inter-convertibility of all forms of energy. Reference: The law of thermodynamics describes the general direction of physical change in the universe. The historical development of the subject is complicated; its origins were early 19th century studies of steam engines by Nicolas Carnot and others, and its main laws were introduced (the first two in their earlier form by Rudolf Clausius in 1850) before heat was understood in terms of the random motion of atoms and molecules. The first law of thermodynamics states the equivalence of heat and work and reaffirms the principles of conservation of energy. The second law states that heat does not of itself pass from a cooler to a hotter body. Another, equivalent, formulation of the second law is that the entropy of a closed system can only increase, that is, the energy in the system will inevitably tend to become distributed in the most probable, i.e. disordered, pattern. The third law states that it is impossible to reduce the temperature of a system to absolute zero in a finite number of operations. Implicit in all this is what is sometimes called the zeroth law, that thermodynamic equilibrium is manifested by equality of temperature. Among areas in which the principles and methods of thermodynamics are of great practical importance are the study and design of engines, and the rules governing the direction of chemical and biochemical reactions (Pearsall and Trumble, 2003).

Thermal: (1) of, for or producing heat, (2) promoting the retention of heat (Pearsall and Trumble, 2003)

The First Law of Thermodynamics: the law of conservation of energy, states that energy can be neither created nor destroyed but can be converted from one form into another. During any physical or chemical reaction the energy form may be changed but the energy output must equal the energy input (Mercuro, Lopez and Preston, 1994).

The Second Law of Thermodynamics: Second law of Thermodynamics: explains what happens to energy when it is used (either convert from one form into another or transferred from one system to another). The second law states that when energy is used it is degraded into a lower quality and less useful form, usually low temperature heat, which is dissipated or dispersed into the surrounding environment. When high-quality energy (concentrated energy with a great potential for doing useful work) is used to perform work some of it will be degraded to low-quality energy or waste heat which has little potential for performing useful work. No conversion of energy into another form or transfer of energy from one system to another is thus 100% efficient (Mercuro, Lopez and Preston, 1994).

Open system: biological systems belong to a class of open systems maintained far from thermodynamic equilibrium through the effect of exchange of energy and material with the outside world; and are characterised by non-linear interaction among the constituent chemical reactions. Such systems typically posses multiple steady states, each stable in some range. This stability assumes the form of homeostasis that tends to maintain the particular steady state or to restore it if perturbed by fluctuations below a certain threshold value. Should perturbations exceed the threshold, the system may jump to an alternative steady state. These systems satisfy the basic requirements of biological regulation as they possess both stability and the capacity for organised change (Ho and Saunders, 1979).

Model: a simplified (often mathematical) description of a system etc., to assist calculations and predications (Pearsall and Trumble, 2003).

Entropy: A brief history of entropy: the concept of entropy first arose in the 19th century as a mathematical quantity in thermodynamics. It was later given a physical interpretation as representing the degree of disorder or randomness of the constituents of physical system, expressed as the probability of occurrence of its particular arrangement of particles. Thermodynamic theory indicates that the entropy of an isolated system can increase but will never decrease (Persall and Trumble, 2003).

Evolution: (1) gradual development, esp. from a simple to a more complex form; (2) Biol. a process by which species develop from earlier forms (3) the appearance or presentation of events etc. in due succession (the evolution of the plot) (4) a change in the disposition of troops or ships (5) the giving off or evolving of gas, heat, etc. (6) an opening out (7) the unfolding of a curve (8) Math. The extraction of a root from any given power (cf. involution) Reference: the philosophical speculation that primitive organisms may change into more complex forms and ascend some imaginary hierarchical ladder is an idea that may be traced back to the ancient Greeks. Jean Bapiste de Lamarck was one of the first to propose a theory of organic evolution, against the prevailing belief in special creation: his suggestion, not now accepted, was that species transform by the inheritance of acquired characteristics. Sir Charles Lyell's demonstration that geological deposits are the product of slow small changes over long periods of time allowed Darwin to develop a theory of evolution by gradual natural selection. In On the Origin of Species (1859) he proposed that an isolated variety within a population might give rise to a new species by survival of the form best suited to its new surroundings; this is what the term "survival of the fittest" means. Most modern explanations of evolution are essentially modifications of Darwin's theory. Combined with recent genetic theory, and the new understanding of the role of DNA. Evolution is the great unifying concept in modern biology (Persall and Trumble, 2003).

Modelling Darwinism evolution: a model of an evolving or learning system. Only a few models of evolution have been constructed to help conceptualise and test the postulates of new-Darwinian theory. Many later simulations of evolution have been attempted for the purpose of improving the adaptation or optimisation process in formal or artificial systems (Casti and Karlqvist, 1986):In the existing models of evolution the environment has been represented as fixed, objective framework that produces the selection pressures on the populations of organisms. one more pedagogic-type model might utilize an artificial environment that could be gradually modified in the hope of inducing new measurements by the organisms; Von Neumann's self-reproducing automation: a genetic description as well as a universal constructor; a set of general conditions or rules of interaction between organism and environment were defined, such as conservation of metabolic resources. The process of natural selection is accomplished by fitness criteria which are explicit and pre-established by the programmer; survival depends on balancing many highly interrelated qualitative system properties such as speed, reliability, efficacy, recovery from error, efficiency and adaptability, perception, action, learning, language, knowledge and other forms of intelligent activity.
Technological barriers of modelling Darwinism evolution: any model of an evolutionary process must clearly represent and functionally distinguish language and measurement constraints (the genotype and phenotype) and must preserve the properties and relations of each. Von Neumann (1966) was more interested in the logical or linguistic

aspects of the model than in the physical aspects of pattern recognition and measurement. Computer languages do not tolerate mutations or re-combinations, whereas genes and natural languages depend on such changes for evolution and creative expression. While some of these dynamic models have helped clarify measurement constraints. Unless an artificial system contains representations of the constraints of both generalised langue and generalised measurement, as well as the complementary relations between them as semantic closure, the model is not likely to evolve similarly to living systems or to contribute significantly to the theory of evolution.

A mathematical model is an idealized theoretical description that reproduces the basic features of a phenomenon, while leaving out its inessential complications (Von Baeyer, 1998).

Parametric design: "detailed design" or "parametric design", where the basic configuration of some component or mechanism is already decided and the purpose is to optimise its performance by adjusting various dimensional parameters. The fitness of solutions is measured using standard simulation software, as for example finite element analysis of structural performance (Steadman, 2008).

Bibliography

Adami, C. (2002) What is Complexity? BioEssays, 24, 1085–1094;

Aon, M. A. and Cortassa, S. (1997) Dynamic Biological Organization: Fundamentals as Applied to Cellular Systems, London; New York, Chapman & Hall.

Allen, T. F. H. (2001) Applying the Principles of Ecological Emergence to Building Design and Construction, in Kibert, C. J., Sendzimir, J. and Guy, G. B. (eds) Construction ecology: nature as the basis for green buildings, New York, Taylor & Francis.

Allenby, B. R. & Graedel, T. E. (1999) Defining the Environmentally Responsible Facility. in Schulze, P. C. (Ed.) Measures of Environmental Performance and Ecosystem Condition. Washington, D.C.: National Academy Press, National Academy of Engineering.

Arnheim, R. (1977) The Dynamics of Architectural Form: Based on the 1975 Mary Duke Biddle Lectures at the Cooper Union, Berkeley, University of California Press.

Atkins, P. and Paula J. D. (2006) Physical Chemistry. Oxford, Oxford University Press.

Atkins P. W. (1984) The Second Law, New York, Scientific American Library: distributed by W. H. Freeman.

Atlan, H. (1974) On the Formal Definition of Organisation. Journal of Theoretical Biology, 45, 295–304.

Azapagic, A. (2004) Life Cycle Thinking and Life Cycle Assessment (LCA), in Azapagic, A., Perdan, S. and Clift, R. (eds), Theory and Practice of Sustainable Development: Case Studies for Engineers and Scientists, John Wiley & Sons, Chichester, pp. 426–437.

Baierlein, R. (1999) Thermal Physics. Cambridge, U.K.; New York, Cambridge University Press.

Bailey, K. D. (1990) Social Entropy Theory, Albany, N.Y., State University of New York Press.

Balian, R. (2004) Entropy, a Protean Concept, Séminaire Poincaré: L'Entropie 06/12/2003 – 06/12/2003 Institute Henri Poincaré, Paris, France 38, Birkhäuser Verlag, Basel, Page. 119–144.

Batty, M. (2006) Visualizing the City, Paper presented at the Post digital Sigradi 2006, Santiago, Chile.

Baxter, R. S., Owers, J. and Echenique, M. (Eds.) (1975) Urban Development Models (LUBFS Conference proceedings), Construction Press.

Behling, S. and Behling, S. (2000) Solar Power – The Evolution of Sustainable Architecture, Prestel Publishing.

Bell, D. (2001) Future of Technology, Pelanduk Pubns Sdn Bhd.

Bellut, C. (2008) "Arh, Luise, lass…das ist ein zu weites feld," or: the Gordian knot of complexity, in Gleiniger, A. and Vrachliotis, G. (eds) Complexity: Design Strategy and World View (Context Architecture), Birkhäuser Basel.

Benyus, J. M. (2002) Biomimicry: Innovation Inspired by Nature, New York, Perennial.

Bergós, J. (1999) Gaudí, the Man and His Work. Boston, Little, Brown and Company.

Bertalanffy, L. V. (1968) General System Theory: Foundations, Development, Applications, New York: G. Braziller.

Birkeland, J. (ed) (2002) Design for Sustainability: A Sourcebook of Integrated, Ecological Solutions. London, Sterling, VA, Earthscan Publications Ltd. (Girardet 1996; Birkeland, 2002).

Birkeland, J. (2008) Positive Development: From Vicious Circles to Virtuous Cycles Through Built Environment Design. Sterling, VA: Earthscan Publications Ltd..

Bisch, J. (2001) Natural Metabolism as the Basis for Intelligent Architecture, in Kibert, C. J., Sendzimir J. and Guy. G. B. (eds) Construction Ecology: Nature as the Basis for Green Buildings, New York, Taylor & Francis.

Boltzmann, L. (1964) Lectures on Gas Theory Berkeley, University of California Press.

Blofeld, J. (1978) Taoism: the Road to Immortality, Boulder, Colo, Shamahala Publications;安乐哲 and 彭国翔 (2004) Ultimate Transformation: Death in Classical Daosim 终极性的转化:古典道家的死亡观. History of Chinese Philosophy 中国哲学史.

Brooks, D. R. and Wiley, E. O, et al. (1988) Evolution as Entropy: Towards a Unified Theory of Biology, Chicago, University of Chicago Press.

Brown, S. K. (1997) Indoor Air Quality, State of the Environment Technical Paper Series, Department of the Environment, Sports and Territories, Australian Federal Government.

Brundtland, G. H. (1987) Our Common Future: The World Commission on Environment and Development, The Assessment of the Brundtland Commission, Oxford; New York, Oxford University Press.

Burley, P. and Foster, J. (eds) (1994) Economics and Thermodynamics: New Perspectives on Economic Analysis, Boston Kluwer Academic Publishers.

Caldas, L., Norford, L., Rocha, J. (2006) An Evolutionary Model for Sustainable Design, Management of Environmental Quality.

Callen, H. B. (1985) Thermodynamics and an Introduction to Thermo-statistics, New York John Wiley.

Carnot, S. (2005) Reflections on the Motive Power of Fire and Other Papers on the Second Law of Thermodynamics, Dover Publications.

Casti, J. L. and Karlqvist, A. (eds.) (1986) Complexity, Language and Life: Mathematical Approaches, Berlin; New York, Springer-Verlag.

Casti, J. L. (1990) Searching for Certainty: What Scientists Can Know About the Future, New York, W. Morrow.

Casti, J. L. (2000) Five More Golden Rules: Knots, Codes, Chaos, and Other Great Theories of 20th Century Mathematics, New York: John Wiley.
Caldarelli, G. (2004) The Structure of Biological and Social Systems, SIAM News, 37.
Carlo, L. D., P. Albisinni, et al. (2006) New Survey Methods for the Construction of Digital Models in an Urban Environment. 3rd International Conference on Virtual City and Territory, Bilbao, Spain.
Carson, R. (1963) Silent Spring. London, Hamilton.
Castells, M. (1989) The Informational City: Information Technology, Economic Restructuring, and the Urban-Regional Process. Oxford, UK; Cambridge, Mass., USA: B. Blackwell.
Castrigiano, D. P. L. and Hayes, S. A. (1993) Catastrophe Theory, Reading, Mass, Addison-Wesley Pub. Co., Advanced Book Program.
Chang, R. (2006) Chemistry. New York, McGraw-Hill Science/Engineering/Math.
Clark, J. O. E. (2004) Thermodynamics: A Dynamical Systems Approach (Princeton Series in Applied Mathematics), Princeton University Press.
Clausius, R. (1850) On the Motive Power of Heat and on the Laws Which Can be Deduced from it for the Theory of Heat. Poggendorff's Annalen der Physick. Poggendoff Annalen, LXXIX (Dover Reprint).
Clausius, R. (1865) Mechanical Theory of Heat: With its Application to the Steam-engine and to the Physical Properties of Bodies, London: Van Voorst, 1867, New York: Readex Microprints, 1968.
Commoner (1972) The Environmental Cost of Economic Growth.
Crawford, R. and Treloar, G. (2005) An Assessment of the Energy and Water Embodied in Commercial Building Construction. 4th Australian LCA Conference, Sydney.
Crutchfield, J. (1994) The Calculi of Emergence: Computation, Dynamics and Induction. Physica D, 75: 11–54.
Cutnell, J. D. and Johnson, K. W. (2001) Physics New York, John Wiley & Sons Inc..
Daintith, J. (2005) A Dictionary of Science. Oxford; New York, Oxford University Press.
Daly, H. E. (1999) Ecological Economics and the Ecology of Economics: Essays in Criticism. Cheltenham, UK; Northampton, MA, USA, E. Elgar.
Daniels, K. (2000) Low-tech, Light-tech, High-tech: Building in the Information Age. Basel; Boston: Birkhäuser.
Davis, S. M. (1986) the Integrated Urban Complex, Unpublished Masters Thesis, College of Architecture and Environmental Design, tempe, Arisona.
Davies, P. (2004) Cosmic Blueprint: New Discoveries in Nature's Creative Ability to Order the Universe, Philadelphia, Templeton Foundation Press.
Delanda, M. (1997) A Thousand Years of Nonlinear History, New York, Zone Books.
DeMarco, T. and Lister T. (1987) Peopleware: Productive Projects and Teams, New York, NY, Dorset House Publishing Company, Incorporated; 2 edition, February 1, 1999.
Dobson, A. (1990) Green Political Thought, Unwin Hyman, London.

Dobzhansky, T. (1956) Genetics of Natural Populations XXV. Genetic Changes in Populations of Drosophila Pseudomonas Drosphila Persimilis in some locations in California. Evolution 10, 82–92.
Docalavich, H. (2007) The UN and the Global Marketplace: Economic Developments. Philadelphia, Mason Crest Publishers.
Durning, A. T. (1992) How Much is Enough? The Consumer Society and the Future of the Earth. London, Earthscan.
Dyer G., China's Data Put it Ahead of Germany, URL: http://www.ftchinese.com/story.php?lang=en&storyid=001024228.
Darwin, C. (1929) The Origin of Species by Means of Natural Selection, London, Watts & Co..
Eddington, A. (1958) The Nature of the Physical World, Ann Arbor: University of Michigan Press, pp.68–80.
Ehrlich, P. R. (1972) The Population Bomb, London, Pan/Ballantine.
Eigen, M. (1971) Self-Organization of Matter and the Evolution of Biological MacroMolecules Naturwissenschaften, 58, 465–523.
Epstein, J. M. and Axtell, R. (1996) Growing Artificial Societies: Social Science from the Bottom up. Washington, DC: Brookings Institution Press: Cambridge: MIT Press, Page 35.
Epstein, J. M. (2006) Generative Social Science: Studies in Agent-Based Computational Modelling, Princeton University Press.
Erve, W. S. v. d. (1951) Le Corbusier, idealistisch Architect Utrecht, Oosthoek.
Faber, M., H. Niemes, et al. (1987) Entropy, Environment, and Resources: An Essay in Physio-Economics, Berlin; New York, Springer-Verlag.
Ferracin, A., Panichelli, E., Benassi, M., DI Nallo, A. and Steindler, C. (1978) Self-Organizing Ability and Living Systems. Biosystems, 10, 307–317.
Frazer, J. (1995) An Evolutionary Architecture (Themes VII), London, Architectural Association Publications.
Frazer, J. (1996) The Groningen Experiment, in P. Johnston (Ed.), Projects Review 1995–96 (pp. 108–111): The Architectural Association School of Architecture.
Friedrich, I. A. and A. I. I+D (2006). The I.T. Energy Consumption Evaluation Tool for the City of the New Millennium. 3rd International Conference on Virtual City and Territory, Bilbao, Spain.
Gans, H. J. (1962) The Urban Villagers: Group and Class in the Life of Italian-Americans, New York: Free Press; London: Collier-Macmillan.
Gauzin-Müller, D. and Favet, N. (2002) Sustainable Architecture and Urbanism: Concepts, Technologies, Examples, Basel; Boston: Birkhauser (Raynsford, 1999: 42; Gauzin-Müller and Favet, 2002).
Georgescu-Roegen, N. (1971) The Entropy Law and the Economic Process. Cambridge, Mass., Harvard University Press.
Glansdorff, P. & Prigogine, I. (1971) Thermodynamic Theory of Structure, Stability and Fluctuations John Wiley & Sons Ltd.

Gleiniger, A. (2008) The Difficult Whole or the Rediscovery of Complexity in aArchitecture, in Gleiniger, A. and Vrachliotis, G. (eds) Complexity: Design Strategy and World View (Context Architecture), Birkhäuser Basel.

Gleiniger, A. and Vrachliotis, G. (eds) (2008) Complexity: Design Strategy and World View (Context Architecture) Birkhäuser Basel.

Grisogono, A. M. (2008) Success and Failure in Adaptation, Land Operations Division, Defence Science and Technology Organisation (DSTO), Edinburgh, South Australia.

Grisogono, A. M. (2008B) Co-Adaptation, Land Operations Division, Defence Science and Technology Organisation (DSTO), Edinburgh, South Australia.

Guy, S. and S. A. Moore (2005) Sustainable Architectures: Cultures and Natures in Europe and North America. New York, Spon Press (Irwin, 2001: 16; Guy and Moore, 2005) (Feeberg, 1999: xiii; Guy and Moore, 2005) (Russel and Williams, 2002: 48; Guy and Moore, 2005) (Sudjic, 1996: 7; Guy and Moore, 2005) (Ross, 1994: 5; Guy and Moore, 2005).

Haddad, W. M., Chellaboina, V. and Nersesov, S. G. (2005) Thermodynamics: A Dynamical Systems Approach (Princeton Series in Applied Mathematics), Princeton University Press.

Hagan, S. (2001) Taking Shape: A New Contract between Architecture and Nature. Oxford, Architectural Press (Shiva, 1993: 78; Hagan, 2001) (Shiva, 1993: 73; Hagan, 2001) (Benton, 1994: 38; Hagan, S. (2001) (Tafuri, 1987: 139; Hagan, 2001) (Ambasz, 1991: 24; Hagan, 2001) (Soper, 1995: 155–56; Hagan, 2001) (Aalto, 1986: 221; Hagan, 2001) (Sassen, 1994: 1; Hagan, 2001).

Hall, C. S. and Nordby, V. J. (1973) A Primer of Jungian Psychology. New York, New American Library.

Hawken, P., Lovins, A. and Lovins et al. (1999) Natural Capitalism: Creating the Next Industrial Revolution, London: Earthscan.

Howard, E., Garden Cities of Tomorrow (1965), edited, with preface by F.J. Osborn ; with an introductory essay by Lewis Mumford. Cambridge, Mass.: M.I.T. Press.

Haynie, D. T. (2008/2001) Biological Thermodynamics Cambridge University Press.

Hemmer, P. C., Holden, H. and Ratkje, S. K. (Eds.) (1996) The Collected Works of Lars Onsager: with Commentary, Singapore; River Edge, N.J., World Scientific;

Hensel, M. & Menges, A. (Eds.) (2006) Morpho-Ecologies, London, Architectural Association.

Hershey, D. (2009) Entropy Theory of Aging Systems: Humans, Corporations and the Universe, London, Imperial College Press.

Herzog, T. (Ed.) (2008) European Charter for Solar Energy in Architecture and Urban Planning Prestel USA.

Hirsh and Smale (1974) Differential Equations, Dynamical Systems and Linear Algebra, Academic Press, New York.

Ho, M. W. and Saunders, P. T. (1979) Beyond Neo-Darwinism, an Epigenetic Approach to Evolution. Journal of Theoretical Biology, 78, 573–591.

Huber, P. & Mills, M. P. (2006) The Bottomless Well: the Twilight of Fuel, the Virtue of Waste, and Why We Will Never Run Out of Energy, New York, BasicBooks.
Hussey, A., (1994) Urban Air Pollution in Megacities of the World, The Geographical Review Article, American Geographical Society.
Imre, L. (1991) Autonomous Energy School for Hungary, Sun at work in Europe 6, 2.
Intergovernmental Panel on Climate Change (1990) Report to IPCC from Working Group 1: Policymakers' Summary of the Scientific Assessment of Climate Change.
Intergovernmental Panel on Climate Change (IPCC) (2007) 4th Assessment Report, "Climate Change 2007: Impacts, Adaptation and Vulnerability.
Ingersoll, R. (1996) Second Nature: On the Social Bond of Ecology and Architecture. Reconstructing Architecture: Critical Discourses and Social Practices. Dutton, T. A. and Mann, L. H. Minneapolis: University of Minnesota Press.
Jacobs, J. (1972) The Death and Life of Great American Cities, Harmondsworth: Penguin.
Jantsch, E. (1980) The Self-Organizing Universe: Scientific and Human Implications of the Emerging Paradigm of Evolution, Pergamon.
Jaynes, E. T. (1957) Information Theory and Statistical Mechanics. Physical review 106, 620–630.
Jaynes, E. T. (2003) Probability Theory: the Logic of Science, New York, Cambridge University Press
(Weldon, 1901; Jaynes, 2003: W. F. R., Biometrika, edited by Karl Person and Walter Frank Raphael Weldon, 1901, Vol 1, page 1, cited by Probability theory, the logic of science, by E. T. Jaynes, 2003).
Jencks, C. (1997) The Architecture of the Jumping Universe: a Polemic: How Complexity Science is Changing Architecture and Culture, London Academy Editions.
Jemtrud, M., P. Nguyen, et al. (2006) A Theory of Artistry for 3D Data Fusion. 3rd International Conference on Virtual City and Territory, Bilao, Spain.
Jin, Y. L. (2005) Daoism, Nature and Man, Life, Reading and Knowledge Triple-Association Press, China, 金岳霖《道，自然与人》自然与人 生活·读书·新知三联书.
Johnson, T. Liu, F., et al.1997, Clear water, blue skies.
Johansen, J. M. (2002) Nanoarchitecture: a New Species of Architecture, New York City, Princeton Architectural Press; Dollens, D. (2005) Digital-botanic architecture, Santa Fe SITES Books; St. Paul, MN.
Jungermann, A. H. (2006) "Entropy and the Shelf Model: A Quantum Physical Approach to a Physical Property." Journal of Chemical Education 83 (1686–1694).
Kay, J. J. (1984) Self-Organisation in Living Systems, Department of Systems Design Engineering. Waterloo, Ontario, Canada, University of Waterloo.
Kay, J. J. (2002) Confusion Between Steady-State, Thermodynamic Equilibrium and Local Equilibrium, in About Some Common Slipups in Applying Prigogine's Minimum Entropy Production Principle to Living Systems.

Katchalsky, A. and Spangler, R. (1968) Dynamics of Membrane Processes. Quart. Rev. Biophys., I, 127.
Katz, I. (2009) Twenty Ideas That Could Save the World, Ingenious, Madcap and Perhaps not Strictly Legal: the Guardian's Search for the Greatest Plan to Tackle Climate Change, www.guardian.co.uk.
Kauffman, S. A. (1993) The Origins of Order: Self-Organisation and Selection in Evolution, New York, Oxford University Press.
Kauffman, S. A. (2000) Investigations, New York, Oxford University Press.
Kestin, J. (ed) (1976) The Second Law of Thermodynamics, Stroudsburg, Pa.: New York (exclusive distributor), Dowden, Hutchinson & Ross, Halsted Press.
Knowles, R. L. (2006) Ritual House: Drawing on Nature's Rhythms for Architecture and Urban Design. Washington, D.C.: Island Press.
Kuhn, T. S. (1970) The Structure of Scientific Revolutions. Chicago, The University of Chicago Press.
Lapthorne, K. (1973) Integration of Autonomous and Industrialised Building Systems Working Paper 7. Cambridge, University of Cambridge, Department of Architecture, Technical Research Division.
Lambert, F. (2002) "Disorder – A Cracked Crutch for Supporting Entropy Discussions." Journal of Chemical Education 79: 187–192.
Langston, C. A. and Ding, G. K. C. (Eds.) (2001) Sustainable Practices in the Built Environment, Boston, Butterworth-Heinemann.
Leach, N. (2009) Swarm Urbanism, in Digital Cities, Architectural Design, Leach, N. (Ed.) (2009) Wiley.
Li, M. and Vitányi, P. (2001) An Introduction to Kolmogorov Complexity and its Application, New York Springer-Verlag.
Li, Z. H. (1986) China's Classical Philosophy and Ideology, Zhongguo gu dai si xiang shi lun, Beijing, Ren min chu ban she: 中国古代思想史论，李泽厚(1986), 中国古代思想史论, 北京: 人民出版社: 新華書店发行.
Lovelock, J. (2006) The Revenge of Gaia: Why the Earth is Fighting Back – and How We can Still Save Humanity. London; New York, Allen Lane.
Lü Junhua, Peter G. Rowe and Zhang Jie, 2001, China Statistical Yearbook 2001, Prestel.
MacKay, D. (2008) Sustainable Energy – Without the Hot Air, Cambridge Publish;
Mackenzie, D. (2007) Mathematics of Climate Change, a New Discipline for an Uncertain Century, Mathematical Sciences Research Institute, Berkeley, CA.
Mainzer, K. (2008) Strategies for Shaping Complexity in Nature, Society and Architecture; in Complexity: Design Strategy and World View (Context Architecture), A. Gleiniger and G. Vrachliotis (eds), Birkhäuser Basel.
Malthus, T. R. (1798) First Essay on Population, London Macmillan, 1966.
Marchettini, N., Pulselli, F. M. and Tiezzi, E. (2006) Entropy and the City, in The Sustainable City IV: Urban Regeneration and Sustainability. Mander, Ü., Brebbia, C. A. & Tiezzi, E. (Eds.) Tallin, Estonia, Southampton, UK; Boston: WIT: 263–272.

Margolis, L. and Robinson, A. (2007) Living Systems: Innovative Materials and Technologies for Landscape Architecture, Liat Margolis, Alexander Robinson, Basel Birkhäuser.
Masanori Shukuya (2000) Design with Natural Potentials, in Ray-Jones, A. (Ed) (2000) Sustainable Architecture in Japan: the Green Buildings of Nikken Sekkei. Chichester, Wiley-Academy.
Mazria, E. (1979) The Passive Solar Energy Book: a Complete Guide to Passive Solar Home, Greenhouse, and Building Design, Emmaus, Pa., Rodale Press; It's the Architecture Stupid!, Solar Today Magazine, May/June 2003; Turning Down the Global Thermostat, Metropolis Magazine, October 2003; Blueprint for Disaster, On Earth Magazine, Summer 2005.
Marx, K. and Engels, F. (1965) Capital: A critical Analysis of Capitalistic Production, Moscow, Progress Publishers.
Marx, K. (1963) The Poverty of Philosophy, New York, International Publishers.
Meyer, W. T. (1983) Energy Economics and Building Design. New York McGraw-Hill.
McDonough, W. and Braungart, M. (2000) Eco-effectiveness: a New Design Strategy, Sustainable Architecture White Papers, New York, Earth Pledge Foundation.
McDonough, W. and Braungart, M. (2002) Cradle to Cradle: Remaking the Way We Make Things, New York, North Point Press.
McHarg, I. L. (1992) Fitness, the Evolutionary Imperative, in Design with Nature, New York, J. Wiley.
McHarg, I. L. (2006) The Essential Ian McHarg: Writings on Design and Nature, Washington, D.C.; London Island Press.
Mclennan, J. F. (2000) Living Buildings, Sustainable Architecture White Papers, New York, Earth Pledge Foundation.
Meadows, D., Meadows, D. and Randers, J. (1992) Beyond the Limits: Confronting Global Collapse, Envisioning a Sustainable Future, Post Mills, Vt.: Chelsea Green Pub. Co..
Meadows, D., Randers, J. and Meadows, D. (2005) Limits to Growth: the 30-year Update, London; Sterling, VA, Earthscan.
Mercuro, N., Lopez, F. A. & Preston, K. P. (1994) Ecology, Law and Economics: the Simple Analytics of Natural Resource and Environmental Economics, Lanham, University Press of America.
Moore, J. W., Stanitski, C. L., et al. (2001) Chemistry: the Molecular Science. U.S. Brooks/Cole.
Moore, P. (2006) Little Book of Big Ideas: Science Sydney, ABC Books for the Australian Broadcasting Corporation.
Moore, S. (2006) Architecture, Esthetics, and Public Health. in The Hand and the Soul: Ethics and Aesthetics in Architecture and Art, Illescu, S. (Ed.) Charlottesville, VA, University of Virginia Press.
Morin, E. (1995). Le vie della complessità. La sfida della complessità. G. Bocchi and M. Ceruti. Milano, Italy, Feltrinelli: 49–60.

Morison, A. W., D. Hes, et al. (2006). Material Selection in Green Buildings and the CH2 Experience. Technique research paper 09. Melbourne.
Naess, A. (1989) Ecology, Community, and Lifestyle: Outline of an Ecosophy. Cambridge; New York: Cambridge University Press.
Nicolis, G. and Prigogine, I. (1977) Self-Organization in Non-Equilibrium Systems: From Dissipative Structures to Order Through Fluctuations, New York: Wiley.
Nicolis, G. and Prigogine, I. (1989) Exploring Complexity: An Introduction, New York W.H. Freeman.
Niklas, K. J. (1978) Theoretical Evolutionary Rates in Plant Groups and the Fossil Record. Brittonia, 29, 241–254.
Norris, C. (2000) Quantum Theory and the Flight From Realism: Philosophical Responses to Quantum Mechanics, London, Routledge.
Odum, H.T. (1971) Environment, Power and Society, New York: Wiley Inter-science.
Odum, H. T. and Odum, E. C. (1981) Energy Basis for Man and Nature, New York, McGraw-Hill.
Odum, H. T. (1983) Systems Ecology: An Introduction. New York: Wiley.
Odum, H. T. (1996) Environmental Accounting, Energy and Environmental Decision Making, New York, Wiley & Sons.
Olgyay, V. and Olgyay, A. (1963) Design with Climate: Bioclimatic Approach to Architectural Regionalism. Princeton, N.J., Princeton University Press.
Oppenheimer, S. (2006) The Origins of the British: A Genetic Detective Story London, Constable and Robinson.
Parnell, M. and Cole, G. (1983) Australian Solar Houses, New South Wales, Second Back Row Press.
Pearsall, J. and Trumble, B. (eds) (2003) Oxford English Reference Dictionary. Oxford English Reference Dictionary. Revised second edition ed. Oxford, Oxford University Press.
Pearsall, J. and Hanks, P. (1998) The New Oxford Dictionary of English. Oxford, Oxford University Press.
Peterson, G. (2001) Using Ecological Dynamics to Move Toward an Adaptive Architecture, in Kibert, C. J., Sendzimir, J. and Guy, G. B (eds) Construction Ecology: Nature as the Basis for Green Buildings, New York, Taylor & Francis.
Pfeiffer, B. (ed) (1939) Frank Lloyd Wright, An Organic Architecture, in Frank Lloyad Wright Collected Writings Vol. 3.
Pike, A. (1974) The Autonomous House Research Programme SRC/DOE Autonomous Housing Study. Science Research Council. Cambridge, University of Cambridge, Department of Architecture, Technical Research Division.
Portoghesi, P. (2000) Nature and Architecture, Milano, Italy, Skira Editore S.P.A..
Prigogine, I. (1961) Introduction to Thermodynamics of Irreversible Processes. New York, Interscience Publishers.
Prigogine, I. (1978) Time, Structure and Fluctuations. Science 201, 777–785.
Prigogine, I. (1980) From Being to Becoming: Time and Complexity in the Physical Sciences, New York, W. H. Freeman.

Prigogine, I. and Stengers, I. (1984) Order Out of Chaos: Man's New Dialogue with Nature, London, Fontana Paperbacks.
Prigogine, I. (1991) The Arrow of Time, in Rossi, C. & Tiezzi, E. (eds.) Ecological Physical Chemistry. Amsterdam, Elsevier, NL.
Pulselli, R. M., Pulselli, F. M., Ratti, C. and Tiezzi, E. (2005) Dissipative Structure for Understanding Cities: Resource Flows and Mobility Patterns. The 1st International Conference on Built Environment Complexity. University of Liverpool.
Pulselli, F. M., Pulselli, R. M., et al. (2006) Environmental Accounting of Buildings: Outcomes from the Emergy Analysis, in Mander, Ü., Brebbia, C. A. and Tiezzi, E. (eds) The Sustainable City IV: Urban Regeneration and Sustainability, Tallin, Estonia, Southampton, UK; Boston: WIT: 489–498.
Pulselli, R. M. and Tiezzi, E. (2009) City Out of Chaos: Urban Self Organization and Sustainability, University of Siena, Italy, WIT Press.
Prokopenko, M., Boschetti, F. and Ryan, A. J. (2007) An Information-Theoretic Primer on Complexity, Self-Organisation and Emergence, at The 8th Understanding Complex Systems Conference, UIUC, 2007.
(Crutchfield and Young, 1989; Prokopenko, Boschetti, and Ryan, 2007) Crutchfield, J. P. and Young, K. Inferring statistical complexity. (1989) Phys. Rev. Lett., 63:105–108, An information-theoretic primer on complexity, self-organisation and emergence, Prokopenko, M., Boschetti, F. and Ryan A. J. (2007) at The 8th Understanding Complex Systems Conference, UIUC, 2007.
Ray-Jones, A., Ed. (2000) Sustainable Architecture in Japan: the Green Buildings of Nikken Sekkei. Chichester, Wiley-Academy.
Rees, W. E. and M. Wackernagel (1996) Our Ecological Footprint: Reducing Human Impact on the Earth. Gabriola Island, BC, New Society Publishers.
Revkin, A. C. "Ian McHarg, 80, Architect Who Valued a Site's National Features", obituary, New York Times, March 9, 2001, page 9.
Roodman, D. M. and Lessen, N. (1995) A Building Revolution: How Ecology and Health Concerns Are Transforming Construction – Worldwatch Paper 124, Worldwatch Institute, Washington, DC.
Rogers, E. M. (1960) Physics for the Inquiring Mind, Princeton: Princeton University Press.
Rogers, R. G. and Gumuchdjian, P. (1998) Cities for a small planet. Boulder, Colo.: Westview.
Rosnay, J. D. (1979) The Macroscope – a New World View. New York, Hagerstown, San Francisco, London Harper & Row, Publishers.
Roy, B. N. (1995) Principles of Modern Thermodynamics Bristol [U.K.], Philadelphia: Institute of Physics Pub.
Ryn, S. V. D. and Cowan, S. (1996) Ecological Design. Washington, DC, Island Press.
Salat, S. (2006) The Sustainable Design Handbook China: High Environmental Quality Cities and Buildings, Hermann Editeurs des Sciences et des Arts (CSTB).
Sahal, D. (1979) A Unified Theory of Self-Organisation. Cybernetics, 9, 127–142.

Sandler, S. I. (2006) Chemical, Biochemical, and Engineering Thermodynamics, Hoboken, N.J., John Wiley & Sons.
Schlesinger, J. R. (1979) Declared Speech to the National Press Club, Washington D.C., August 16, Congressional Record 9/24/79.
Schneider, E. D. and Sagan, D. (2005) Into the Cool: Energy Flow, Thermodynamics, and Life, Chicago, University of Chicago Press (Lotka, 1945, 194; Schneider and Sagan, 2005).
Schrödinger, E. (1962) What is Life?: the Physical Aspect of the Living Cell, Cambridge, England, Cambridge University Press.
Schroeder, D. V. (1999) An Introduction to Thermal Physics Addison Wesley.
Schumacher, E. F. (1973) Small is Beautiful: a Study of Economics as if People Mattered, London, Blond and Briggs.
Schumaker, P. (2006) We Think This is a New Paradigm for Designing. Architecture Journal, 21(12), 40.
Shannon, C. E. (1949) The Mathematical Theory of Communication. Urbana, University of Illinois Press.
Shelden, D. (2009) Feel the force, Closing the Gap: Information Models in Garber, R. (Ed.) Contemporary Design Practice, Hoboken, N.J.: Wiley; Chichester: John Wiley.
Singleton, D. (2004) Sustainability, a Risk Management Perspective, Arup, London, UK.
Stasinopoulos, P., Smith, M. H. et al. (2008) Whole System Design: An Integrated Approach to Sustainable Engineering. London, UK, Earthscan Ltd..
Steadman, P. (2008) The Evolution of Designs: Biological Analogy in Architecture and the Applied Arts, Routledge.
Steele, J. (1997) Sustainable Architecture: Principles, Paradigms, and Case Studies, New York, McGraw-Hill.
Steele, J. (2005) Ecological Architecture: A Critical History 1900-Today, London, Thames & Hudson.
Stonier, T. (1990) Information and the Internal Structure of the Universe: an Exploration into Information Physics. London; New York: Springer-Verlag.
Stotland, A., Pomeransky, A. A., et al. (2004) The Information Entropy of Quantum Mechanical States, in Europhysics Letters 67, 700.
Stowe, K. (2007) An Introduction to Thermodynamics and Statistical Mechanics, Cambridge, UK; New York Cambridge University Press.
Smil, V. (1999) Energies: an Illustrated Guide to the Biosphere and Civilization (MIT, 1999), Page xv, Table 3.
Terzidis, K. (2008) Algorithmic Complexity: Out of Nowhere, in Complexity-design strategy and world view (Context Architecture), Gleiniger, A. and Vrachliotis, G. (eds), Birkhäuser Basel.
Thirring, H. (1976) Energy for Man: from Windmills to Nuclear Power, New York, Harper & Row. Indiana University Press, page 157.

United States Green Building Council's Leadership in Energy and Environmental Design (LEED)

United Nations (2006) World Urbanization Prospects: The 2005 Revision and Carl Haub (Population bulletin, world population highlights, key findings from PRB's world population data sheet, 2007).

United Nations (2009) The Millennium Development Goals Report 2009. New York, United Nations Department of Economic and Social Affairs.

Vale, B. and Vale, R. J. D. (2000) The New Autonomous House: Design and Planning for Sustainability, London: Thames & Hudson.

Vallero, D. and Brasier, C. (2008) Sustainable Design: the Science of Sustainability and Green Engineering, Hoboken, N.J., John Wiley.

Venturi R. (1950) Context in Architectural Composition, Excerpts from M.F.A thesis, Princeton University, in Complexity-design strategy and world view (Context Architecture), (2008) Gleiniger, A. and Vrachliotis, G. (eds), Birkhäuser Basel.

Venturi, R. (1988) Complexity and Contradiction in Architecture, London; Sydney, Butterworth Architecture.

Von Baeyer, H. C. (1998) Maxwell's Demon: Why Warmth Disperses and Time Passes, New York, Random House.

Von Neumann, J. (1955) Mathematical Foundations of Quantum Mechanics. Princeton, N.J., Princeton University Press.

Von Neumann, J. (1966) Theory of Self-Reproducing Automata, Urbana, University of Illinois Press.

Vrachliotis, G. (2008) Popper's Mosquito Swarm: Architecture, Cybernetics, and the Operationalization of Complexity, in Gleiniger, A. and Vrachliotis, G. (eds) Complexity: Design Strategy and World View (Context Architecture), Birkhäuser Basel.

Walker, S. (2006) Sustainable by Design: Explorations in Theory and Practice, London; Sterling, VA, Earthscan.

Walker, S. (2007) Gordon Matta-Clark: Matter, Materiality, Entropy, Alchemy. In Thomas, K. L. (ed.) Material Matters: Architecture and Material Practice. London ; New York, Rutledge Taylor & Francis Group.

Watson, J. and Crick, F. (1953) "Molecular Structure of Nucleic Acids: A Structure for Deoxyribose Nucleic Acid", Nature 171, no.4356, 25 April 1953: 737–738.

Watson, J. D. and F. H. C. Crick (2005) Molecular Structure of Nucleic Acids: a Structure for Deoxyribose Nucleic Acid. Research Advances in Genetics and Genomics: Implications for Psychiatry. N. C. Andreasen. Washington, D.C.; London, American Psychiatric.

Wicken, J. S. (1978) Information Transformations in Molecular Evolution. Theoretical Biology , 72, 191–204.

Wiener, N. Congress, N. W. C. (1997) Proceedings of the Norbert Wiener Centenary Congress, 1994: Michigan State University, November 27-December 3, 1994, Providence, R.I., American Mathematical Society.

Wines, J. (1987) De-Architecture, New York, NY, Rizzoli.

Wines, J. (2000) Green Architecture. Köln; London, Taschen.
Wines, J. (2000B) The Art of Architecture in the Age of Ecology, Sustainable Architecture White Papers, New York, Earth Pledge Foundation.
Wright, R. (2000) Non-Zero: the Logic of Human Destiny, New York, Pantheon.
Yeang, K. (1995) Designing with Nature: the Ecological Basis for Architectural Design. New York:, McGraw-Hill.
Yeh, H. (2003) A Study on the Meaning & Measurement Model Building up for Application of Theory of Entropy in Internal Auditing, Chung Yuan Christian University.
World Wide Fund for Nature.
The World Challenge, Finance Time, by Editorial, December 15, 2006.
www.arctic.noaa.gov
www.bfi.org/introduction_to_bmf.htm, Buckminster Fuller Institute
www.cap-lmu.de Global Statistics, Research Group on the Global Future, July 2005.
www.inhabitat.com
www.thewashingtonnote.com
www.nrri.umn.edu
http://nobelprize.org/nobel_prizes/economics/laureates/2002/public.html
www.santafe.edu, Emergence, Organisation and Dynamics of Living Systems, Santa www.time.com
http://www.complexsystems.net.au/wiki/Complex_%2709_Workshop_on_Causality_in_Complex_Systems Complex '09 Workshop on Causality in Complex Systems, COSNet.
www.simplexconf.net Simplex 2009, the 1st Annual Workshop on Simplifying Complex Network for Practitioners.
Hwww.simplexconf.net Simplex 2009, the 1st Annual Workshop on Simplifying Complex Network for Practitioners.
(2000) Emission Database for Global Atmospheric Research.
(2000) Sustainable architecture white papers, New York, Earth Pledge Foundation.
(2004) McGraw-Hill concise encyclopaedia of chemistry. New York, McGraw-Hill.
(2004) United Nations Human Settlement Programme, The State of World Cities: Globalisation and Urban Culture, New York.
(2005) International Earth Science Information Network.
(2005) Catalogue: Foster and Partners. Munich; London, Prestel.
(2007) 2007 Human Development Report (HDR), United Nations Development Program, November.
Lester Brown, President of the World-Watch Institute.
U.S. Energy Information Administration.

All the websites were last checked in May 2011.

Index

Appendix: A diagram of transitional phases of open systems evolution

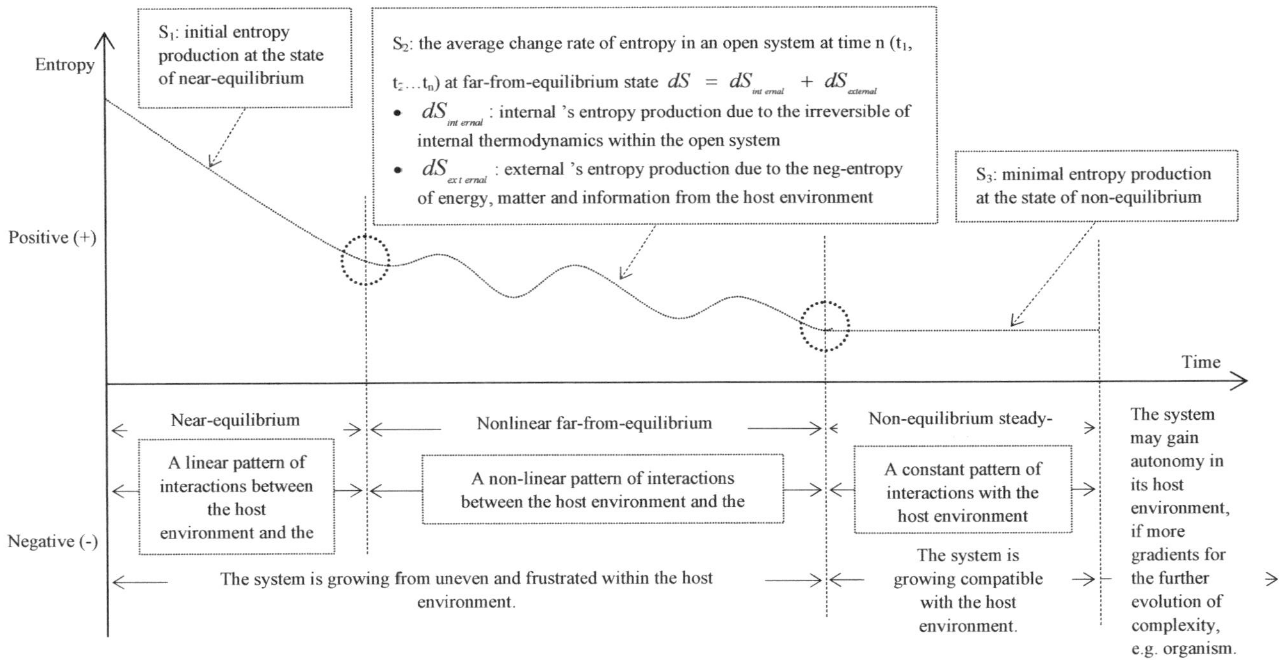

FSC
www.fsc.org
MIX
Papier | Fördert gute Waldnutzung
FSC® C083411

Kolibri 360 GmbH
Ferdinand-Jühlke-Straße 7
99095 Erfurt, Deutschland
produktsicherheit@kolibri360.de